KU-323-573

INTERIBERICA, S.A. DE EDICIONES

FRONTIERS OF LIFE

Animals of Mountains and Poles

Doubleday and Company Inc.,
Garden City, New York, 1976

SBN: 385–11346–3
Library of Congress Catalog Card No: 75–13114

Also published in parts as Polar Life and Mountain Life.

FRONTIERS OF LIFE

Part 1
Polar Life

by Joseph Lucas
and Susan Hayes

ISBN: 84-382-0020-6. Dep. Legal: M. 221-1976.
Printed and bound in Spain by Novograph S.A., and Roner S.A., Crta de Irun, Km.12,450, Madrid–34

Series Coordinator Geoffrey Rogers
Series Art Director Frank Fry
Design Consultant Guenther Radtke
Editorial Consultant David Lambert
Series Consultant Malcolm Ross-Macdonald
Art Editor Douglas Sneddon
Editors Maureen Cartwright
Damian Grint
Research Barbara Fraser
Art Assistant Michael Turner

Contents: Part 1

Foreword by David Attenborough

Cold is the enemy of life. At its extreme, it is lethal. It can freeze fish and reptiles solid. Birds and mammals, which must maintain their body temperature within a small range, may quickly become chilled beyond recovery. Even the hardiest of plants can be killed by its savage bite. The greatest ally of cold is the wind. Blizzards can suck away warmth faster than even the best insulated animal can maintain it. Gales strip plants of their leaves and rip all but the most firmly-rooted from the ground. And where high winds and low temperatures operate together, they create the most hostile environments on earth—the mountains and the poles.

On the highest mountains, where scarcity of oxygen adds a further oppression to the wind and cold, there is virtually no life whatever. Yet a little below those sterile peaks, plants, amazingly, manage to grow. Lichens coat boulders. Hugging the rock, their leaves clasped close to their stems, tiny bushes cling to crevices. A little lower still, are the specialized communities that are often unique to particular ranges, sometimes even to individual mountains. Peaks that rise from low-lying plains may be as isolated from one another as islands in an ocean. As a consequence, the creatures that live on them may evolve into unique forms just as animals do on lonely islands such as the Galápagos. Bernard Stonehouse provides a fascinating guide to these mountain floras and faunas—the extraordinary thickets of giant groundsels on the flanks of Mount Kenya, the high-altitude hummingbirds of the Andes, and the giant pandas and golden takins of the mountains of western China.

The temperatures at the poles are lower than on any but the highest mountains. A few mammals—polar bears and seals—can be found on the ice that floats over the North Pole, but the South Pole, placed as it is in the center of a huge continent, is almost entirely devoid of life—except, of course, for that irrepressible colonizer, man. The seas that surround Antarctica, however, are rich with microscopic floating creatures and these form the pastures that, directly, or indirectly, support huge numbers of creatures—fish, seals, whales, penguins, and albatrosses. Indeed, these creatures have solved the problems of existence here so well that the penguin rookeries and sea-elephant herds are among the most spectacular assemblages of animals in the world. And, as Joseph Lucas and Susan Hayes, our guides to the polar regions say, these are probably "the last truly wild places left on earth."

Throughout the following pages, the authors describe, again and again, the elaborate mechanisms that creatures have developed in order to survive the savage, killing cold. The wonder, perhaps, lies not only in the ingenuity of these adaptations, but in the seemingly unquenchable urge that has driven animals to colonize these most inhospitable of regions, the ends of the earth, the very frontiers of life.

David Attenborough.

The Ends of the Earth

Most of us mentally picture the polar regions as bitterly cold, snow- and ice-covered, blizzard-racked places at the northern and southern ends of the earth. We know that the polar bear somehow makes its home in the north and that penguins thrive in the south. But we probably feel that no man in his right mind would want to go to these inhospitable regions. For thousands of years, though, people have actually lived in the north polar lands fringing the Arctic Ocean, lands that now hold perhaps half a million human inhabitants. Even the remote and once uninhabited Antarctic continent at the southern end of the earth today has a year-round population. About 3000 people live there during the summer, when tourists arrive by the shipload.

In recent years man's activities in polar regions have brought both of these frigid areas almost as much attention as they received in the early 1900s, when explorer-adventurers heroically raced to be first at the poles—the northernmost and southernmost points on the earth, for which the surrounding polar regions are named. In 1969 the giant tanker *Manhattan* proved that the so-called Northwest Passage through the Canadian part of the Arctic Ocean was indeed navigable, although it is doubtful whether this route around North America will ever become commercially useful. Less glamorously, oil spills in the cold Arctic waters have spread their slimy and tarry blight along stretches of coastline, and on land the heavy vehicles of oil-exploration companies have inflicted long-lasting damage on the fragile polar plant-soil system known as the *tundra*. Antarctica received a brief blaze of publicity during the International Geophysical Year, especially when the Commonwealth Trans-Antarctic Expedition crossed that continent in 1957–8. Some 10 years later the first tourist cruise ship, *Lindblad Explorer*, nosed its way into waters west of the Antarctic Peninsula.

In spite of the publicity given to such events,

The 1007-foot-long American tanker Manhattan *has an especially strengthened icebreaker bow 69 feet in length. She forced the Northwest Passage in 1969. On her return voyage, the tanker was slightly damaged by the ice, highlighting the possibility of oil spills on a massive scale in polar waters.*

most of us still have a rather hazy notion of what the polar regions comprise. Let us start, then, by defining them. Broadly speaking, the Arctic consists of the Arctic Ocean bordered by many islands and two major landmasses—northern Eurasia and North America. The Antarctic comprises a massive central comma-shaped continent surrounded by the southern waters of the Atlantic, Indian, and Pacific oceans—often regarded as forming one Southern Ocean where they surround Antarctica.

It is easier to describe the polar regions in such general terms than to fix their boundaries precisely. As far as geographers are concerned, the north polar region, or Arctic, is the earth's surface north of the Arctic Circle, an imaginary line that girdles the earth at 66° 33′ North. This is the Northern Hemisphere's southernmost latitude to receive a period of at least 24 hours of continuous daylight in summer and at least 24 hours of unbroken darkness in winter. (At the North Pole each period lasts roughly six months.) Similarly, geographers define the south polar region, or Antarctic, as the earth's surface south of the Antarctic Circle at 66° 33′ South. Here, of course, periods of midnight sunshine and mid-day darkness fall at exactly opposite times of the year from those in the north.

But maps showing the Arctic and Antarctic as perfectly symmetrical regions bounded by circles can be misleading, because polar climate, polar ocean currents, and polar plants and animals observe no such simple limits. For each region a more realistic natural boundary might be the farthest limit of the mass of more or less continuous ice on the sea, but this varies from season to season, and even from year to year. The iceberg limit fluctuates even more widely. What the naturalist needs is some dividing line that will give a good index of polar conditions

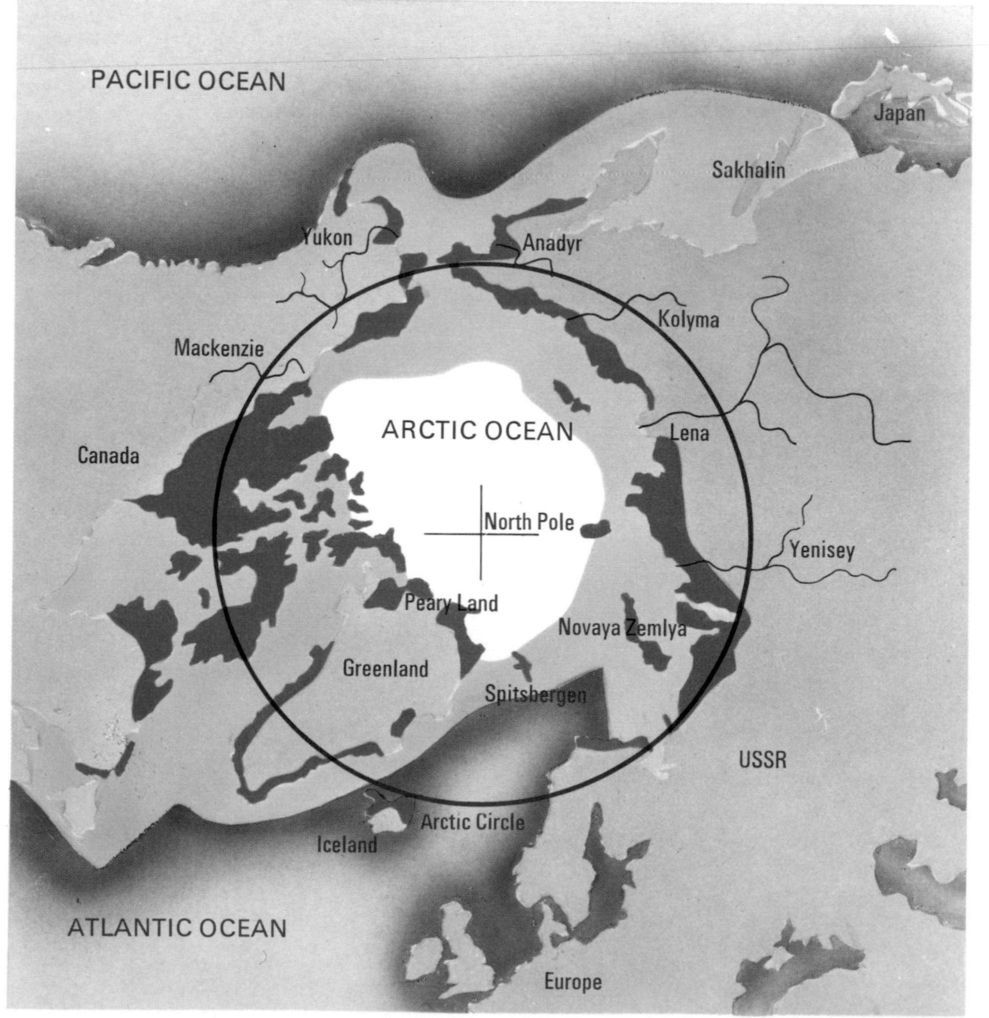

This map of the Arctic shows that the Arctic Circle serves as a misleading guide to the area covered by Arctic conditions. At sea, permanent ice occurs only well inside the circle. In winter, pack ice extends far to the south of the circle off the northeast coasts of Asia and North America, yet only drift ice reaches seas south of Spitsbergen. On land, the treeless tundra zone begins north of the Arctic Circle in much of northern Asia and parts of North America, though it starts south of the circle in northeast America and extreme northeast Asia. Year-round ice grips the mountains of Greenland far to the south of the circle. Local variations in the southern extent of tundra and pack ice around the North Pole reflect local variations in temperature. Ocean currents in the Atlantic, and major rivers such as the Mackenzie bring fingers of warmth to places that would otherwise have to endure much colder climates.

unaffected by short-term fluctuations.

In the Northern Hemisphere the limit that best serves this purpose is the July 50°F isotherm—an imaginary line joining places on the earth's surface where the July temperature reaches but does not exceed 50°F. Where this line passes over land it roughly coincides with the northern limit for tree growth. (Even where winters are bitterly cold, trees will grow provided the summer temperature is sufficiently high.) North of the rather abrupt timberline, the shrubs and sedges of the treeless tundra extend to the stony, barren, high Arctic and to the Arctic Ocean except where ice- and snow-clad mountains and glaciers stand. South of the timberline, and thus outside the true Arctic, lies the huge tract of mainly coniferous forest known as the *taiga*. Here the chief trees are spruce and birch in North America, pines in northern Scandinavia (where the trees almost reach the shores of the Arctic Ocean), and various conifers in northern Russia and Siberia.

Defining the limit of the south polar regions raises its own problems. The obviously polar coast of the Antarctic continent extends well north of the Antarctic Circle in many places, but the coast is not a climatic boundary, because the cooling effect of the Southern Ocean carries the polar climate far north of the mainland. Here, then, the generally accepted polar limit is an oceanographic one—the junction where cold, Antarctic surface waters meet the warmer waters of the north. This so-called Antarctic Convergence is about 20 to 30 miles wide and encircles the entire Southern Ocean. Its average position varies, but lies at roughly 50° South through most of the Atlantic and Indian oceans, moving north to 55–62° South in the Pacific.

Clearly the two polar regions offer considerable contrasts. This becomes especially plain

Limit of pack ice in March

Limit of pack ice in September

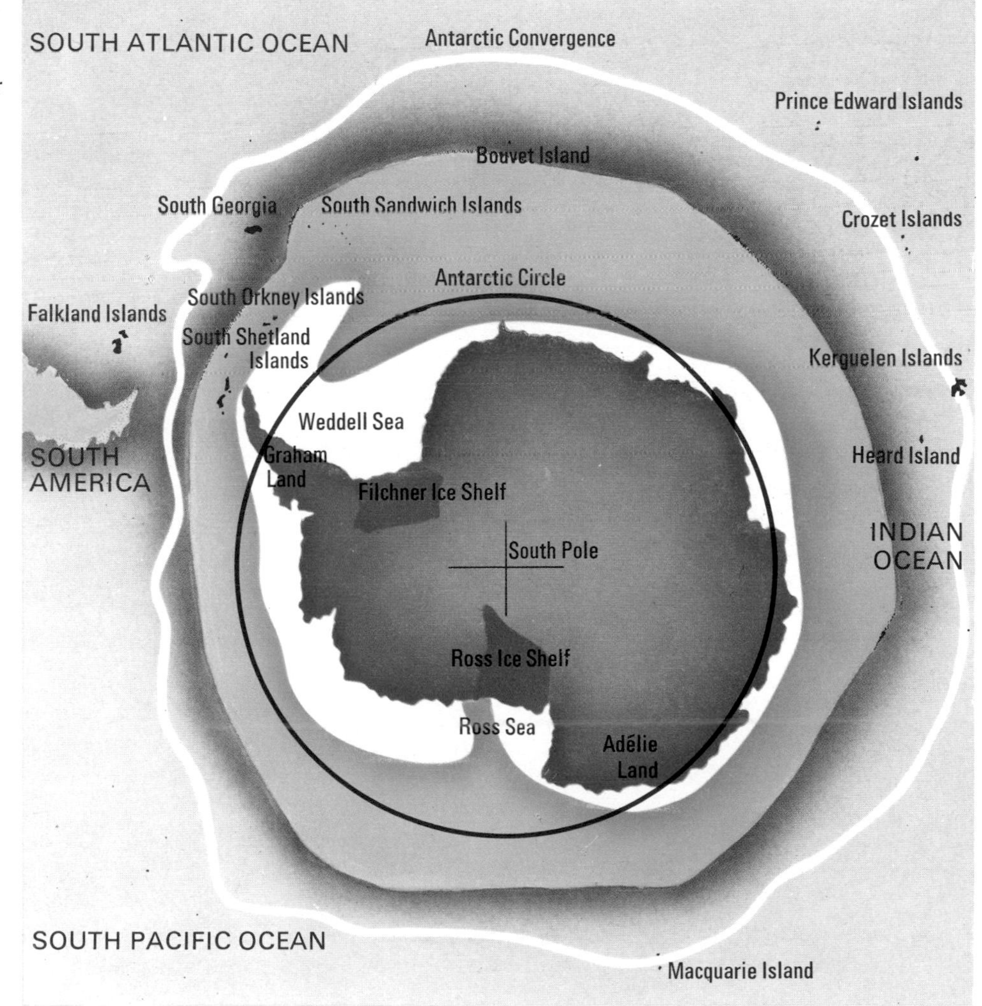

Unlike the Arctic, all the land and sea inside the Antarctic Circle must face the rigors of a polar climate. Here there are no fingers of warmth to penetrate the polar circle. Instead, the intensely cold air from the high, central Antarctic continent flows downward and outward, chilling the continental coasts and their surrounding seas. Thus, although this map covers as much of the world as the Arctic regions map on the facing page, because of the extremely harsh climate it includes no land with trees. Instead, low-growing grasses, mosses, and lichens clothe favored parts of oceanic islands and mainland coasts. Almost all other land lies beneath everlasting snow and ice. Permanent ice shelves jut far out from Antarctica itself, and fringing seas are usually locked in pack ice. The true rim of the Antarctic is the Antarctic Convergence, where the cold Antarctic ocean water sinks below warm water from the north.

Glaciers sweep down to the coast on the spectacularly beautiful island of South Georgia. They form high up in the mountains and slide slowly downhill, their enormous weight gouging out U-shaped valleys. Icebergs break off at the seaward end.

when we compare their land surfaces. Antarctica, the continent that forms the hub of the south polar region, occupies about five and a half million square miles, which makes it roughly as big as the United States of America and Mexico combined, or two and a half times as large as Europe. Antarctica is the highest continent on earth, with an average altitude of 6000 feet. But 95 per cent of its surface is permanently covered by an ice sheet up to 12,000 feet thick that blankets plains, mountains, and valleys alike. Except around the coast, the only ice-free land consists of *nunataks*—isolated mountain peaks that poke up through the ice sheet. As the ice accumulated over many thousands of years, it gradually spread outward from Antarctica's interior in the form of huge glaciers that eventually reached the sea. Here, at many places, ice masses now flow out over the sea surface to form vast floating shelves. The two largest of these are the Ross Ice Shelf, which is roughly as big as France, and the combined Ronne and Filchner ice shelves, which exceed Italy and Switzerland put together. From time to time chunks of the seaward edges of these shelves break off under the influence of the tides or of swells originating in storms far to the north, and form flat-topped icebergs up to 1000 feet thick and occasionally 100 miles or more across.

Unlike Antarctica, northern lands as a whole bear no massive burden of ice and they lack permanent ice shelves. A great ice cap indeed covers much of Greenland, but elsewhere permanent ice usually persists in the local form of valley glaciers—in Alaska, for instance. (When they reach the sea, such glaciers break up into craggy icebergs, and not into the flat-topped type of icebergs formed from the Antarctic ice shelves.) However, ice sheets did once cover much of the Arctic. There they ground down many mountain ranges, so that a great deal of the far north is now low-lying land.

Both the Arctic and Southern oceans carry sea ice that increases in area in winter and spring. Antarctic sea ice reaches its greatest extent in

The Ross Ice Shelf, in east Antarctica, covers an area of 210,000 square miles. It has been formed by the outward movement of ice and snow from the interior of the continent, and its 1000 feet or more of thickness are actually floating in the sea. In the early days of Antarctic exploration it was called the Great Ice Barrier because it blocked the way to the coast; it was discovered in 1841.

September or October, when it vastly enlarges the ice shelves that surround Antarctica, effectively doubling the continent's area. Antarctic sea ice shrinks to its minimum area in March, when ships can sometimes approach parts of Antarctica's shores unimpeded. In the Arctic, however, sea ice always covers a vast part of the Arctic Ocean surrounding the North Pole.

Both ends of the earth share broadly the same type of intensely cold climate. This is because each is tilted away from the sun in winter and only slightly toward it in summer. Thus for a time in the winter each region receives no sunlight at all, and in the summer the sun's radiation arrives only very obliquely, providing far less heat than the much more direct radiation that beats down on the tropics. Other factors reinforce the cooling effects caused by lack of hot sunshine. For instance, cloud largely covers the Arctic and the Southern Ocean, reflecting a good deal of the incoming solar radiation before this reaches the earth's surface. Yet more radiation is reflected by the snow and ice cover. Then, too, much of the solar radiation that is absorbed is largely used up in melting a small part of the snow and ice.

Although both polar regions have a cold climate they are not equally cold. Antarctica is even icier than the Arctic. This is partly because Antarctica is higher, and temperature drops with a rise in altitude. The coldest known place on earth stands in Antarctica. This is near the Soviet research station *Vostok* (South) which is located over 10,000 feet up, at the "Pole of Cold" on the high plateau of eastern Antarctica. Vostok's recorded minimum temperature of −126.4°F (more than 158 degrees of frost) is far below the Northern Hemisphere's lowest known temperatures of −89.9°F at Verkhoyansk in Siberia and −81°F at Snag in the Yukon.

Taken on its own, the high average altitude of Antarctica would make this region some 22°F colder than the Arctic, but in fact the overall temperature difference between the two places is greater than that, for the Arctic is warmer than its high latitudes would suggest. This relative warmness is due to the moderating influence of the Arctic ocean, which, although largely under an icy "lid," receives warm waters moving northward from the Atlantic and Pacific oceans. Besides bringing low-grade heat to the Arctic Ocean these waters also warm its surrounding shores. An example of temperature differences

Tabular bergs off the Antarctic coast (left) are enormous chunks of ice that have broken from the shelves under the influence of tides or ocean swells. These flat-topped icebergs may be 1000 feet or more in thickness; the largest known was 208 miles long and 60 miles wide, an area greater than Belgium. Pack ice is formed by the freezing of the sea in winter and is usually up to nine feet thick. During the warmer weather of summer the floes spread out to give the open pack ice shown (above foreground).

influenced respectively by cold and warm oceans is the fact that South Georgia at 54° South has a permanent ice cap and glaciers, whereas Liverpool in England and Hamburg in Germany, both near 54° North, experience temperate climates.

A feature of great importance to all living organisms in polar regions is the wind. Besides being the coldest place on earth, Antarctica is also the windiest, especially that part known as Adélie Land. Throughout the Southern Ocean generally the winds blow from the west, forever circling the world in the latitude belts aptly nicknamed the Roaring Forties, Furious Fifties, and Screaming Sixties. Close to the Antarctic coast, however, winds subjected to the influence of local disturbances blow from the east. These disturbances are often linked with masses of cold, dense air forming high on the Antarctic plateau. The relatively heavy air masses slide down the continental slope, accelerating as they go. Such *katabatic* winds may reach the coasts as savage gales gusting at speeds of more than 90 mph. Curiously, perhaps, the awesome Antarctic blizzards or snowstorms raised by these winds depend not upon heavy snowfall but upon the power of the gale to lift lying snow from the surface and to whip its hard, granular, abrasive crystals horizontally through the air. For example, a 10-mph wind is enough to start granules of dry snow moving across the surface, and at 20 mph the drift is high and dense. When the wind speed rises to 30 mph the pain caused by the driven spicules striking uncovered skin is intense, and at 40 mph traveling is almost impossible because visibility is reduced to near zero.

In the mainly low-lying Arctic there is no comparably large katabatic effect, and the wind blows less strongly and less frequently than in the southern regions. Nevertheless, Arctic storms are just as unpleasant, and the "wind chill" makes an already cold environment seem even colder. A fine, sunny, windless day in polar regions can be very pleasant even though the temperature is below zero, but at the rise of a breath of wind the apparent temperature drops suddenly. As the wind speed increases, the sensation of cold gets more pronounced and physiological effects become more severe. Under such conditions the risk of frostbite is very real. Because of his height man is more exposed to the cold winds than most

polar animals. Arctic foxes and seals, living close to the ground, can find sheltering windbreaks where the wind has blown hollows in the snow, and such creatures can sleep peacefully as drifting snow forms an insulating quilt.

Despite the snow they receive, it is perhaps surprising to find that both polar regions are technically deserts. This is partly because snow is a dry form of water, although sleet and even rain have been recorded on polar fringe areas. Then, too, not a great deal of snow falls in a year. Actual measurements are unreliable because it is impossible to distinguish between falling and drifting snow. But scientists can estimate snowfall at a particular place if they have measured snow accumulation and evaporation there over a period of years. It seems that between 12 and 24 inches of snow fall annually in Antarctica. Some of this blows out to sea, a little evaporates, and the rest is compressed into a layer of ice four to eight inches thick and weighing up to two trillion tons. The Arctic's snowfall is at much the same annual rate, but relatively more snow melts and evaporates here, so accumulation is correspondingly less. The Greenland ice cap—the Arctic's largest—is only about one eighth the size and one tenth the weight of the ice cap that covers Antarctica. It is nevertheless capable of producing from 10,000 to 15,000 icebergs each year. No icebergs are calved from Peary Land, in the northern part of Greenland, however, for although it lies a mere 500 miles from the North Pole its annual precipitation equals only about one inch of rain, which is less than many tropical deserts receive, and not enough to build glaciers sufficiently large to yield icebergs.

It seems that neither polar region has been perpetually ice-covered, but then Antarctica at least has not always been where it now is. Scientists believe that hundreds of millions of years ago all the continents were joined together as one supercontinent, now known as Pangaea.

Many scientists believe that about 420 million years ago, in the Silurian period, the super-

Left: Adélie Land, in east Antarctica, is one of the windiest places on earth. Masses of cold dense air, forming high on the Antarctic plateau, begin to slide downhill toward the coast. Funneled and concentrated by the shape of the land, the winds may accelerate to speeds of at least 90 mph. Animals, such as the Adélie penguins below, sit out the accompanying fierce blizzards.

continent surrounded the South Pole, which at that time was more or less in what is now mid-Atlantic. Gradually the mass drifted over the surface of the globe and broke in two: a great northern continent called Laurasia, comprising North America and Eurasia (minus the Indian subcontinent), and a great southern continent called Gondwanaland, comprising South America, Africa, the Indian subcontinent, Australia, New Zealand, and Antarctica. Land bridges allowing the migration of plants and animals between the two masses must for a while have been present. Eventually, both the giant continents also broke up. Gaps opened up between Labrador and Greenland, and North America rotated away from Africa, creating the Atlantic Ocean. The break between Greenland and Europe may have happened no more than 6 million years ago, in the Eocene period. The components of Gondwanaland seem to have separated about 115 million years ago, during the Cretaceous period, although Australia and Antarctica do not appear to have split until about 6 million years ago.

There is still a good deal of uncertainty about precisely when and where the continents wandered: for instance, one school of thought places Antarctica in mid-Pacific at one time. But there is plenty of proof that Antarctica was formerly linked with the other big southern landmasses and India. For instance, fossils reveal that all these regions once featured closely related warm-climate plants dominated by two distinct genera, *Glossopteris* and *Gangamopteris*, not found in the more northern lands of Laurasia. Also, telltale marks in rocks help to show that some 290 million years ago, before Gondwanaland broke up, a great glaciation spread across what are now southern Africa, South America, India, parts of Australia, and Antarctica. More fossil evidence linking these lands dates from after this glaciation and includes the remains of closely related mammal-like reptiles that lived here in warm conditions about 200 million years ago. Some of this geological evidence proves that Antarctica at times had a subtropical, if not tropical, climate. What caused Antarctica's subsequent decline into a further glacial phase is unknown, but the climate undoubtedly cooled. There were warm interludes, especially in the Cretaceous period about 135 to 65 million years ago, but by about 5 million years ago the present ice cap had formed.

The recent ice age in the Arctic was conceivably triggered by, and probably began after, that in the south. Scientists comparing ancient layers of ice from both regions are still studying possible correlations between the phases of glaciations in the Arctic and the Antarctic, but it now seems that the last glaciation reached its peak of severity at about the same time in both places.

We have seen that over millions of years the polar regions have become broadly alike but by no means identical. Both are dry deserts with short, cool summers and long, cold winters. But climates and land areas differ in various ways, and these contrasts influence the relative distributions of life forms. In the Arctic, the central surface is hostile to terrestrial life, for it consists of a mass of sea ice. Here, the only food for animals is found in the sea beneath the ice. The high Arctic therefore supports virtually no wholly or partly land-based life except where seals, seabirds, and seal-hunting polar bears can profit from breaks in the ice. Terrestrial life on the Antarctic plateau is even more scanty, and largely restricted to the seabirds called skuas, occasionally seen near the South Pole, and to scattered growths of lichens colonized by tiny invertebrates and found on isolated mountain peaks that pierce the blanket of ice and snow.

The subglacial tundras surrounding these icy heartlands appear superficially alike in both hemispheres. All tundra features permanently frozen subsoil (*permafrost*) that impairs drainage and produces waterlogged conditions when the surface thaws in summer. But the Arctic tundra is much more extensive and contains many more species of plants and animals than the south.

In spite of such differences, the same general problems face polar organisms, whether they inhabit the Arctic or the Antarctic. In lands that are mainly cold, dark, and lacking liquid water for a large part of the year, the organisms must somehow obtain fresh water and food, capture and retain, or manufacture, enough heat, and in most cases compress the vital reproductive parts of their life cycles into the short summer season. Creatures unable actively to survive the severe polar winters must be able to hibernate or to migrate. Our next chapters help to show just how plants and animals cope with "life in an icebox."

"Midnight sun" shines over Greenland during the summer. The limit of 24-hour daylight in summer and 24-hour darkness in winter are the Arctic and Antarctic circles. The periods of summer daylight and winter darkness lengthen toward the poles.

Plants and Invertebrates

Polar lands that seem dead in winter may teem with life in summer, for, somehow, a variety of living things has colonized the cold ends of the earth. Among these hardy organisms are certain land plants and invertebrates (many of them vital foods for other forms of land-based polar life). This chapter explores these linked ideas. In particular, it explains how polar plants and invertebrates cope with cold and how those in the north differ from their southern counterparts.

During their long winter nights, the polar surfaces appear almost lifeless, except for the activities of certain animals. In the north, polar bears roam the ice, feeding on seals that rise in breathing holes. Arctic foxes follow the bears at a respectful distance, scavenging from their kills, and an occasional snowy owl flies overhead in search of any lemming unwary enough to leave its tunnel under the snow. In the south, almost the only sign of life outside the sea comes from the breeding colonies of emperor penguins watching over their eggs and chicks in the darkness. In the surrounding ocean, Weddell seals chew breathing holes in the ice cover and perhaps listen to the songs of the Ross seals. While some mammals and birds are active on the polar surface during winter, the land plants and invertebrates of both polar regions remain motionless beneath a snow mantle.

In spring and summer, however, a transformation occurs. Life returns at least to parts of the polar regions in an abundance of individuals, if not of species. The Arctic tundra glows with flowers, buzzes with insects, and abounds in birds that make their nests in the far north. For the brief summer season, reproduction and growth are usually hectic, though not invariably so. A bad summer season can prevent the appearance of, or destroy, most young plants and animals. Fortunately, many Arctic organisms have long enough breeding lives to make good such losses.

The summer wealth of the Antarctic cannot match that of the Arctic, but in the far south, grass, moss, and lichen begin their slow growth, and small invertebrates creep and jump in sheltered nooks among the vegetation. Migrating seabirds now return to Antarctica in their

During the very brief Arctic summer the whole tundra glows with flowers whose bright colors attract passing insects. Plant growth has to be rapid, because flowers must bloom and seeds mature before the onset of the severe winter. Many Arctic plants grow close to the ground, to take advantage of the warmer air just above the earth's surface and to avoid the worst of the wind and weather.

Above: stone rings or frost polygons, seen here in Spitsbergen, are the result of repeated cycles of freezing and thawing, which cause expansion and contraction of the surface. These seasonal disturbances heave stones to the surface and segregate them into geometrical patterns covering many square miles, where little grows.

millions, and seals haul themselves out on the breeding beaches and the ice floes. The air vibrates to shrieks, cries, and groans as adult birds pair off, dispute nesting sites, and challenge predators, and as their young cry out for food. The bellowing roars of bull elephant seals reverberate as these bulky mammals battle for territory and harems.

Because permanent snow and ice made almost all of both polar regions uninhabitable until the relatively recent past, one might well wonder how they acquired the plants and animals just described. To find out, let us take each region in turn—first, the Arctic.

Wildlife immigration into the Arctic followed two kinds of event in earth history: changes in the areas of land and sea, and changes in the area of land under ice.

The last great advance of Arctic ice, which was at its peak around 15,000 years ago, helped to lock up so much water in the form of ice sheets that the level of the oceans fell, at one time by 300 feet, revealing the beds of shallow areas of sea. Among the lands exposed was the part of the continental shelf that now forms the submerged floor of the Bering Strait between Asia and North America. The resulting land bridge was as wide as Alaska is now and it remained dry for thousands of years. The Bering land bridge had a subarctic climate, and the evidence of prehistoric pollen suggests that it was covered with a sedgy and grassy tundra. Meanwhile, much of neighboring northeast Asia remained free from glaciers, and ice-free corridors of land linked this area with the relatively warm steppes and forests farther south. Plants and animals from these Asian regions could thus migrate eastward across the Bering land bridge into northern Alaska, which was also evidently ice-free. For a while, the Asian invaders were probably prevented from moving much deeper inside North America by an icy mass south of the Alaska

Mosses and lichens are the earliest colonizers of the harsh polar environment, and even flourish on bare rock and stones. The trunk of a dead tree (above) may become thickly covered with these slow-growing plants. Lichens are extremely resistant to very low temperatures and to desiccation, and each consists of a partnership between an alga and a fungus. Crustose lichens (below right) grow very slowly and some are believed to be 4000 years old, whereas foliose forms (below left) grow relatively rapidly.

Range and another one farther east, in the Mackenzie River area.

When large-scale melting began about 10,000 years ago, the sea again covered the Bering Strait, halting the overland flow of plants and animals into North America. But in Eurasia and North America alike, living things began to colonize land bared by the shrinking of the ice sheets. Kinds of plants and animals that had been pushed south by advancing ice accordingly began moving northwards as the ice retreated. Thus, the North American Arctic has been relatively recently repopulated with a mixture of species of plants and animals—some derived from Asian immigrants, some from original inhabitants pushed south by ice or trapped as relict populations in ice-free "oases" or refuges. In Arctic Eurasia, on the other hand, all plants and animals derive from kinds that merely moved northward back to their ancestors' homelands.

Many of the animals that recolonized Arctic North America and Eurasia after the ice sheets melted quickly died out there, some perhaps pushed into extinction by Stone Age hunters. The victims included mammoths, bison, and horses—indeed, some 95 per cent of all big animals. Many of these forms were plant-eaters, and the disappearance of these herbivores and their predators must have greatly affected Arctic plant communities in ways we do not yet fully understand.

In contrast to the Arctic regions, Antarctica has not been linked by land to warmer areas in recent times. Thus Antarctica's land-based plant

and animal inhabitants probably all derive from ancestors that arrived by sea or air—seals and penguins swam there, and seeds, spores, lightweight invertebrates, and flying birds were borne in on the wind. Once land plants and animals had arrived in the polar regions, their survival there depended on a variety of factors, including availability of water and nutrients, temperature range, length of day, cycle of seasons, and the nature of the surface that they lived on. Polar conditions are so testing that only few plants have proved able to survive them well. For instance, much of the land exposed in the high Arctic consists of bare rock and stones, arguably the harshest of all types of terrestrial habitat. The only plants able to survive on such surfaces are lichens and a few mosses. Lichens are in fact the product of two different plants living in close association for mutual benefit, a type of biological partnership called *symbiosis*. One of the partners is a fungus, the other is a green or blue-green alga. Each has some advantages that the other lacks.

Fungi are plants whose "bodies" often take the form of slender threadlike filaments called *hyphae* that intertwine to produce a web or mat called a *mycelium*. Fungi cannot synthesize food from simple inorganic nutrient salts, water, and carbon dioxide, as green plants can. Instead, they live in or on dead or living plant and animal material. Through the walls of the hyphae, fungi secrete enzymes that decompose this food material. The fungi also absorb water and nutrients through the hyphal walls. The hyphal threads cling to life tenaciously, surviving extremes of temperature and considerable drying.

Algae generally tolerate heat and drought less well than fungi, but, unlike fungi, they can manufacture the compounds that they need for growth. An alga and a fungus that join forces as a lichen can therefore pool their resources and thus survive harsh climatic conditions, including the severe cold and dehydration found in polar regions. Joining forces starts when a fungal spore germinates near enough to the right kind of alga for the hyphae to surround the alga and incorporate it into the mycelium. Here the alga grows and multiplies, and somehow influences the growth of the hyphae. Instead of a delicate network, the result is a much tougher encrusting or "leafy" structure called a *thallus*. Within this thallus both plant types normally live in harmony. The fungus absorbs and stores large quantities of water that the alga needs for food production, and excretes nitrogenous material and phosphates that provides the alga with a source of nourishment. In return, the fungus acquires some of the carbohydrate food manufactured by the alga.

There are altogether some 17,000 known kinds of lichen in the world, hundreds of them found in polar regions. The common rock-dwelling Arctic lichens take one of two general forms: *crustose* or *foliose*. Some crustose types grow as red, orange, or yellow blotches and smears on rock,

Great areas of tundra may become waterlogged in the warmer months, as here in Alaska. Because of low snowfall, the tundra is climatologically desert, but the permanently frozen subsoil prevents meltwater from draining away when the surface thaws.

the so-called "stains of time." Other crustose lichens are considerably larger and thicker, forming warty patches on the surface. All crustose lichens are firmly fixed by means of hyphae that penetrate cracks in rock, and cannot be removed, even with a knife blade, except as specks and flakes.

By contrast, the foliose ("leafy") lichens are attached to rock only at intervals, or indeed only at a single point. The thallus is thicker and has more lobes than that of crustose forms, and it may be borne more or less upright, so that these lichens rather resemble small seaweeds. They are aggressive invaders and often grow relatively fast, crowding out any crustose lichens sharing the rock that they have colonized. But most lichens grow extremely slowly. Some crustose forms annually increase in radius only by between four thousandths and two fifths of an inch. Scientists estimate some individuals to be more than 4000 years old, which would make these plants among the oldest living organisms.

Some scientists have suggested that lack of soil, or the slow rate at which it is formed, may be more important than low temperature in accounting for the stunted and low-growing types of vegetation found in the Arctic. Weathering and the action of lichens in rotting rocks produce a coarse, mineral soil that collects in rock crevices and pockets where lichen debris and perhaps some windborne material enrich it. Pioneers among the higher plants then take root and increase the soil's humus content, and the first invertebrates move in. But the whole process of Arctic soil formation proceeds at a snail's pace compared with soil formation in warmer lands.

In the subarctic and tundra regions, snow and ice blanketed the soil for many thousands of years. Even now, although much of the soil is free from ice, the subsoil remains perpetually frozen *(permafrost)* wherever the temperature averages below freezing for the whole year. During warmer months the surface thaws to a variable depth. This thickness is the *active layer*. The lowest level reached by summer thaw is the *permafrost table*. The known thickness of the permafrost layer below this level varies from a few feet at the southern margins of the Arctic to 1400 feet at Resolute on Cornwallis Island in the Canadian Arctic, but experts believe that permafrost attains a thickness of at least 2000 feet in places. Because meltwater cannot penetrate below the active layer, in summer this becomes a quagmire, restricting plant species to those that can tolerate extremely wet conditions. The waterlogging produces poorly aerated, acid soils where organic decay by bacterial action proceeds extremely slowly, and nitrogeneous and other nutrient salts required by plants are often lacking, except close to the breeding places of birds and mammals, and where dying animals have fallen. In such places nutrients are often plentiful and plant growth tends to be particularly lush and green. Moreover, because plant roots cannot penetrate the permafrost layer there are no deep-rooted plants in the Arctic; instead, many plants anchor and nourish themselves by means of horizontally extensive root systems. Incidentally, such roots often have an oval cross section because most of their annual growth occurs on the upper side, away from the permafrost layer.

There is thus the paradox that although climatologists rank the Arctic as a desert, swamps cover most of its level ground in summer. Of course, if the underlying permafrost thawed, the surface meltwater would drain away and never be completely replaced. Indeed, where the land slopes sufficiently to permit runoff, cold desert conditions prevail all the year round. During the winter, liquid fresh water is available nowhere except in a few hot springs, but Arctic plants manage to survive without it, either in a state of dormancy or by overwintering as seeds. Certain kinds of lower plants, such as the algae, lichens, and mosses, appear to suffer no harm from being frozen. Naturally they become brittle, but when the thaw arrives they resume their normal, slow way of life. Some seeds at least have been shown to be incredibly resistant to prolonged cold: for instance, an Ottawa research station has grown Arctic lupins from seeds that reputedly lay buried in permanently frozen soil on Canada's northwest Arctic coast for more than 10,000 years. Indeed, some scientists believe that seeds dormant since the beginning of the Ice Age could still prove viable.

For those parts of Arctic plants above the soil the main adverse factors are the intense cold and the wind. Plants suffer no less than animals

The stems and leaves of many tundra plants are thickly covered with silky hairs to help reduce the water loss. These hairs also tend to keep the plant quite a lot warmer because they trap an insulating layer of still air close to its surface.

from the effects of wind-chill—the apparent drop in temperature that accompanies a rise in windspeed. In conditions of extreme cold, the fluids in the cells of certain plants may freeze and, as they freeze, expand, bursting the cell walls just as freezing water bursts water pipes. But many Arctic plant species have become able to prevent this happening by developing frost-resistant sap containing dissolved substances that act in the same way as the antifreeze compounds put in automobile radiators in the winter.

Arctic plants are adapted to the cold in other ways as well: for instance, they tend to hug the ground. They cannot grow to any great height partly because winter winds whip up snow, sand grains, and gravel fragments, and these travel fast enough to damage and destroy the delicate growing points of any vegetation projecting above the snow cover. The winds also have a drying effect severe enough to kill exposed stems and leaves. Low-growing plants survive such potential disasters. As a further protection against windblown particles and wind-caused desiccation, the overwintering buds of many plants form just below the surface of the soil or lie hidden among the persistent leaves, leafstalks, or stipules (leaf appendages) of former years. By growing close to the ground, Arctic plants are also able to lessen the effects of wind-chill, because the speed of the wind drops dramatically within a few inches of the surface. At the same time, by growing close together, shrubby plants form a mass of vegetation four or five inches high, resembling (and in places actually composed of) a forest of miniature trees. This further curbs the force of the wind, and because vegetation and the dark soil absorb the long-wavelength infrared rays from the sun, the herbage contains a microclimate that may be considerably warmer than the air immediately

The powder-puff flowering heads of Arctic cotton grass growing in a frozen Alaskan pond (left) bob and sway with every passing breeze. Several kinds of these perennial sedges grow in areas where the tundra is damp in summer. Tundra valleys are usually thickly carpeted with various kinds of grasses (above), and here and there the colorful Arctic poppy sometimes blooms (below).

above. On a May day in northern Greenland, for example, when the noon air temperature stood at 10°F, scientists found that it was 38°F among the dry leaves of a single saxifrage plant, and 50°F in the heart of a clump of moss.

Most tundra-dwelling plants are adapted to drought and indeed share certain drought-resistant characteristics with plants growing in much warmer deserts. To reduce water loss, for instance, the surfaces of some kinds of leaves are covered with a thick, often waxy, waterproof cuticle, and many types of leaves, stalks, and stems are clothed in a dense felting of hairs. By trapping a stationary layer of air close to the surface of the plants, these hairs also tend to keep them warmer than the often very cold surrounding atmosphere.

Perhaps because there are relatively few herbivorous animals to devour them, plants of the polar regions lack the defensive features found in plants of temperate and tropical zones. For example, no polar plants have stinging cells, prickles, or thorns. Any plant that looks appetizing can be eaten safely, because there are no poisonous mushrooms, roots, leaves, or berries.

Arctic plants are designed to withstand cli-

Left: the prostrate and straggling branches of the dwarf willow rise only a few inches above the surface in exposed sites, but they may be up to 20 feet in length. In sheltered valleys, however, these trees occasionally reach six feet or more in height.

matic rigors rather than assault by animals. They are also well equipped to take advantage of the meager blessings of their polar setting. The plants not only survive intense cold, but actually grow rapidly at temperatures low enough to keep most vegetation almost at a standstill. Growth in green plants depends upon photosynthesis, the process in which the green substance chlorophyll helps to produce carbohydrates from water and carbon dioxide in the presence of sunlight. Photosynthesis does not normally start until the temperature exceeds about 43°F. Above this threshold, the life processes of plants speed up as the temperature rises. The rate of this acceleration is far greater in polar regions than in warmer areas: for example, a plant located in the Arctic speeds up its growth rate much more rapidly when the temperature rises from 43°F to 53°F than a plant located in the tropics when the temperature rises from 80°F to 90°F, although in both cases the temperature climbs by 10 degrees. Consequently, many Arctic plants can grow and mature within a shorter season than their relations in kindlier regions. But within the Arctic, patterns of growth vary with aspect. South-facing slopes generally receive more sunshine and are thus warmer than other types of terrain, and they tend to support more plants and more kinds of plants than comparable areas of level ground or north-facing slopes.

The distribution of some plants, however, depends upon light as much as temperature. Many Arctic plants require long spells of daylight, or even continuous daylight, in order to produce flowers and ripe seeds. Even if the cold were less intense, an Arctic winter is too dimly lit to encourage plants to grow. For most of the long winter the only light comes from the softly glowing moon and the stars, brilliantly glittering and reflected as millions of points of light from the snow crystals on the ground. Sometimes the colored, flickering curtains of light of the Northern Lights or aurora borealis appear high in the

Left: reindeer moss is really a kind of lichen, and forms a valuable source of winter food for the reindeer (or caribou, as the same species of animal is called in North America). The lichen grows on sandy or gravelly soils to a height of two or three inches.

The yellow flowers of the perennial mountain avens (above) and the blooms of the purple saxifrage (below) are typical of plants that rely on insects for pollination. Both these plants form compact cushions that break the force of the wind and create a milder microclimate in which the plants can flourish.

sky, as subatomic particles strike the earth's upper atmosphere. But all such sources of light are too weak to set the process of photosynthesis in motion. Any green plants living in the high Arctic must "sleep" through a night that lasts from fall to early spring, and as far south as the Arctic Circle the winter sun peeps above the horizon for only an hour or so each day, and not at all in midwinter.

It is the combination of changes in light, heat, and the availability of moisture that sets plants growing in the Arctic spring. Gradually, the amount of light increases, temperature rises, snow recedes, and the top few inches of the soil thaw. Seeds now germinate rapidly, and many plants that overwintered in a dormant state sprout new leaves and burst into flower.

Most polar plants are perennials, spreading their growth and reproduction over several seasons. Few annuals grow in the high Arctic, and it is easy to see why. Such plants have to cram growth, flowering, and seed production into a single season. After that the parent plants die, leaving only their seeds to carry on their kind from one year to the next. If a poor summer prevented a species of annual from producing seeds, that species would be wiped out. The high Arctic has brief and uncertain summers, and few species of annuals can survive there. Indeed even among the more numerous perennials seed production would not always ensure survival. Many seed-bearing perennials solve this problem by also propagating vegetatively. Some, for example, produce underground stems, called *rhizomes*, containing reserves of food. These stems radiate from the parent plant and send up new aerial shoots some distance away. Other plants throw out runners: long slender stems that grow along the surface of the ground and put down roots at intervals. The resulting plantlets become independent when the stems linking them to the parent plant die.

During the summer, many of the less climatically harsh parts of the tundra give a general impression of a limitless expanse of green grasses, sedges, and rushes up to about a foot high, gently waving in the breeze. The green is broken at intervals by the nodding white heads of cotton grasses (actually sedges) and the golden flowers of the Arctic poppy. Here and there, however, grow shrubs such as the creeping birch, dwarf willow, and Lapland rosebay, which have developed the art of hugging the ground to a fine degree. Their branches—up to 20 feet long in some birch trees—lie almost prostrate upon the surface, nowhere rising more than a few inches above it. And yet these same types of plants, when growing in sheltered valleys, lift their branches as much as six feet high, and appear far more like the trees they really are. On sandy or gravelly soils, extensive areas are covered with reindeer moss, which is not a moss in fact, but a lichen that forms a valuable source of winter food for reindeer (or caribou, as they are called in North America).

Where conditions are harsher, and particularly on stony ground, the so-called cushion plants become commoner. Mosses frequently take this form, and among the herbaceous plants the mountain avens is a typical example. As an adult plant it appears as a small, compact, green mound of leaves topped by delicate-looking white or pale-cream flowers. In the case of the purple saxifrage the young plant starts its growth as a straggly ground-hugger: the young stems grow out horizontally from a stumpy central stem, and send up short flowering shoots at intervals. Gradually, however, it thickens up to form a cushion a few inches across. This cushion shape serves the individual plant in much the same way as the miniature forests that we mentioned earlier serve plant communities: cushions and forests both break the force of the wind and create relatively mild microclimates in which plants can flourish.

One of the most striking features of large tracts of tundra in summer is the number and variety of flowers. In some areas, for a brief period, this otherwise inhospitable land blazes with radiant color. Obviously this pageant is not there to satisfy the aesthetic taste of the occa-

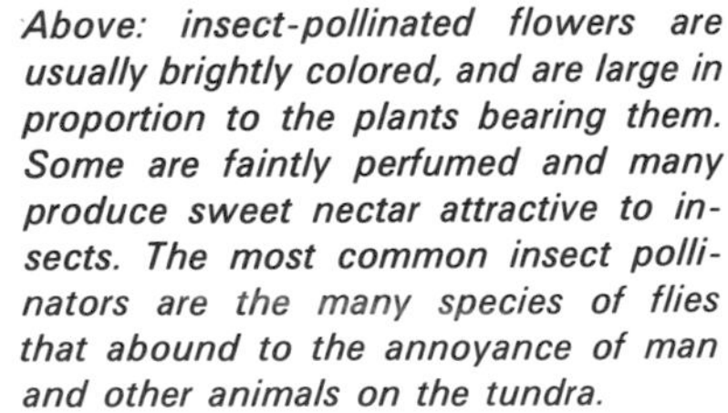

Above: insect-pollinated flowers are usually brightly colored, and are large in proportion to the plants bearing them. Some are faintly perfumed and many produce sweet nectar attractive to insects. The most common insect pollinators are the many species of flies that abound to the annoyance of man and other animals on the tundra.

Left: catkins of the dwarf willow produce very large quantities of ripe pollen, which are shed as the catkins bob about in the wind. Brightly colored flowers are not necessary to the plant, because the wind usually carries some of the pollen by chance to the female flowers.

Spindly-legged crane flies (left) and hover flies (photographed in flight, above) are among the many insects living on the Arctic tundra. The mosquitoes are the most painfully obvious flies of the region, but are probably less abundant than the root maggot flies and midges, which, however, do not bite man. The larvae of many species of Arctic flies either require damp conditions or are truly aquatic in their habits.

sional wandering caribou, Arctic fox, or human being; its purpose is to attract pollinating animals to ensure that seeds are fertilized. But not all flowers are equally colorful: for instance, those of willows and birches are an unassuming green or yellow. These plants have no need to attract animal pollinators, because their seeds are pollinated by the wind. Catkins bobbing and swinging in the wind bear anthers that shed huge quantities of ripe pollen grains. The wind bears these tiny grains away, and by chance a very few land on the stigmas of female willow or birch flowers to fertilize their seeds. Such a hit-or-miss method of pollination obviously works, but it is also wasteful, because only one in many millions of pollen grains will actually fertilize a seed. Some Arctic plants greatly reduce the amount of pollen required by using a self-pollinating mechanism in which pollen-producing stamens and seed-producing ovaries are situated together on the same plant. Again brightly colored flowers are unnecessary, and flowers of self-pollinators are often remarkably unobtrusive.

It is among the plants relying on insect pollination that the most attractive flowers are found. Most Arctic flowers are red, orange, yellow, or blue, and relatively large compared with the size of the plants that bear them. Certain colors attract certain kinds of insect, but the colors and some other features of Arctic flowers serve as more than just visual signals. For instance, the lack of white flowers in the Arctic is connected

Fragile-looking butterflies, such as this swallowtail (left), flit from flower to flower close to the ground. Their larvae feed during the summer and fall, and in the winter hibernate among the vegetation. Some hibernate in fully exposed locations, however, and these larvae probably contain an antifreeze. Springtails can even survive being frozen into ice (below), and are found farther north than other insects.

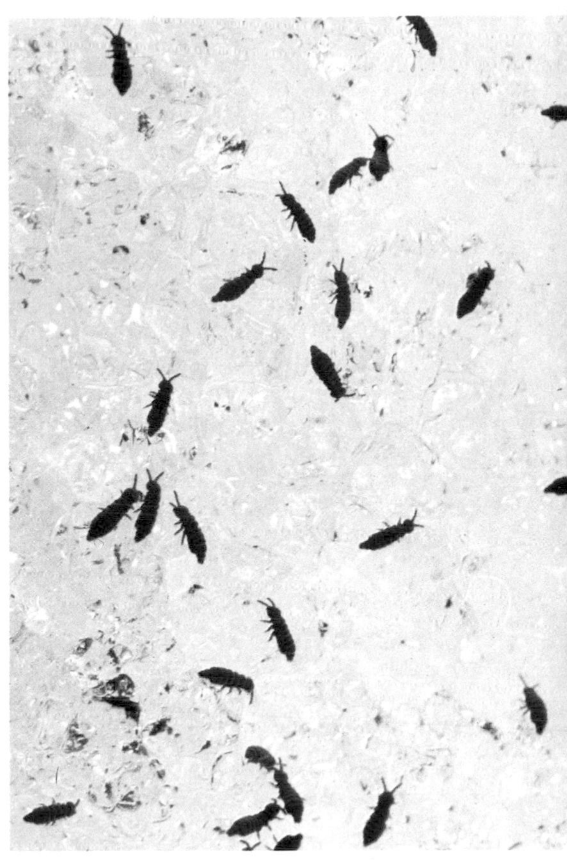

with the low level of solar radiation there. Dark petals absorb solar energy better than pale ones, and may consequently be several degrees warmer than the surrounding air. Dark flowers are thus commoner than pale ones on the tundra. Moreover many flowers are shaped like deep cups, a feature helping to concentrate the available heat. A few Arctic flowers are even heliotropic, turning their blossoms always toward the sun, like the sunflower of warmer regions. But the main purpose of most Arctic flowers remains insect attraction, and to that end many are not only brightly colored but also faintly perfumed, and possess glands called *nectaries* that produce sweet nectar attractive to their visitors.

The commonest insect pollinators in the Arctic are the many species of flies that abound to the considerable annoyance of man and other animals. Round-bodied bumblebees are also important, pollinating as they go from flower to flower, gathering the nectar from which they make a small amount of honey. Male and worker bumblebees die at the onset of winter, only the queen surviving to start a new colony in the spring. Incredibly, in the far north there are also fragile-looking butterflies that flit from one flower to another, close to the ground, to take advantage of the relative warmth.

Like the plants that furnish many of them with their food, Arctic invertebrates are generally well equipped for life near the North Pole, although at first glance the butterflies that we have just mentioned appear an absurd exception to this rule. Their caterpillars feed throughout the summer and fall, then some hibernate among the vegetation or just beneath the surface of the soil; but others set out on a seemingly suicidal course. When the first winter winds begin to blow, they move to exposed ridges and prepare for dormancy in a crevice, beneath a stone, or attached to a clump of vegetation. Here they bear the full brunt of the icy, drying winds and temperatures that fall as low as −70°F in midwinter. How the caterpillars survive without freezing solid remains something of a mystery, but it is possible that they convert glycogen—a starchlike substance in their bodies—into glycerol, an antifreeze. Be that as it may, the ridges upon which they have endured the winter are among the earliest places to warm up at the onset of spring, and the first places where plant growth begins anew. Warmed by the sun, the caterpillars emerge from their nine-month torpor and begin to feed again as though nothing untoward had happened. The insects then pupate, spend a brief time as chrysalids, and reappear as adult butterflies to mate and lay their eggs.

There are about 20 kinds of butterflies in the Canadian Arctic and many more in the extreme north of Europe and Asia. There is also a considerable number of moths, but unlike those of temperate regions these tend to be day-flying creatures with small eyes. They are generally small, and can fly strongly against a stiff breeze.

By far the most abundant insects in the Arctic are the flies. Mosquitoes are especially notorious because of the trouble they cause man and other animals, but they probably rank third in order of abundance, after the midges and root-maggot flies. Male and female mosquitoes both feed on plant juices, the males especially on nectar, but the females need a meal of mammalian blood before they lay their eggs. Like mosquitoes, most midges produce eggs that hatch into aquatic larvae, but unlike mosquitoes, midges do not bite man. Other Arctic insects include the crane flies, or daddy longlegs, hover flies, blowflies, and bumblebees. Parasitic wasps that live inside the larval forms of butterflies, moths, flies, and sawflies are also numerous. Most are small, inconspicuous insects, but as a group these wasps are vitally important because they control the numbers of the insects that they feed on.

Insects that depend on green plants for food do not occur as far north in the Arctic as those able to subsist on lichens, plant debris, and the remains of animals. Such lowly forms include collembolans, or springtails—minute insects, about one tenth of an inch long, that creep within the relatively warm microclimates of the cushions and rosettes of moss and lichen to feed upon decaying plant material. These curious and primitive flightless insects have an ancestry that scientists have traced back directly to the oldest known fossil insects, living about 320 million years ago during Devonian times. Springtails are appropriately named from the sudden leaps they can make when disturbed, thanks to a special type of mechanism beneath the rear of their bodies. Here, an embryonic pair of legs has become modified into a *furca* (springing organ). At rest, this folds forward against the underside of the abdomen, where it is gripped by small "teeth." A sudden contraction of the abdominal body wall muscles sharply increases the pressure of the body fluids within the furca, and the "teeth" holding the organ simultaneously release their grip. The furca then flicks downward away from the body, and propels the insect forward as it strikes the ground. This method of jumping is quite different from that of other insects, such as the flea and the grasshopper, which use their hind legs to the same effect.

Springtails have a very wide distribution over the land surface of the earth, living in all but the most arid places and those where ice and snow never melt. They occur in fresh water, too, and between the tide marks along the seashore, where each high tide submerges them. In the Antarctic, these little creatures have been found from sea level up to 6000 feet at 77° South, where the temperature can fall as low as −85°F. Scientists have even discovered them within six degrees of the South Pole. In both polar regions springtails seem to be more numerous among mosses than among lichens, whereas mites tend to favor lichens as their habitat. Mites—minute creatures related to spiders—have proved to be particularly tolerant of low temperatures: some live on lichen-encrusted rocky peaks within 300 miles (about five degrees) of the South Pole.

In contrast to the north polar regions, though, the Antarctic is poorly endowed with plants and invertebrates. Antarctica itself has no trees or shrubs; it is the only continent totally devoid of them. The Southern Hemisphere's timberline lies far away to the north, at 54° South, along the northern shore of the Beagle Channel in Tierra del Fuego. Here, some 700 miles north of the northernmost tip of the Antarctic Peninsula, the nearest trees to Antarctica are the wind-blasted southern beeches of southern Patagonia.

We can get a broad idea of the poverty of Antarctic plant life in general by comparing the plant types in the polar regions. At one extreme stands the vegetation of the Arctic island group of Spitsbergen. Its climate is softened by the northern tentacles of the Gulf Stream (or, more strictly, of that current's extension called the North Atlantic Drift), and Spitsbergen boasts almost 150 species of flowering plants. Even Arctic North America and Eurasia harbor more than 100 species of such plants, and an abundance of mosses and lichens. In Antarctica the commonest plants are lichens (about 400 species), followed by freshwater and terrestrial algae (about 360 species). Although 70 different kinds of mosses have been recorded here, botanists have identified no more than two species of flowering plants—a grass, and a pink, both of which may be recent arrivals from South America. They have colonized a number of subantarctic islands (early-19th-century sealers perhaps introduced the grass to the South Shetland and South Sandwich islands) but their only roothold on the polar continent is down the west coast of the Antarctic Peninsula to Marguerite Bay (about 64° 35′ South). Nowhere on the mainland are they plentiful, and—unlike those from the islands—continental specimens are stunted and rarely flower.

People sometimes call the region where these flowering plants live the Maritime Antarctic. This includes the west coast of the Antarctic Peninsula, the Palmer Archipelago, and the South Shetland, South Orkney, and South Sandwich islands. All these places have a climate influenced by the nearby ocean. At least one summer month has a mean air temperature above freezing point at sea level, and the mean monthly temperature in winter rarely falls below 14°F. Moreover, there is a moderate amount of precipitation, some falling as rain in the summer. Conversely, in the Continental Antarctic, precipitation is meager and never falls as rain; furthermore, average monthly air temperature at sea level never rises above freezing point and in winter it falls below −4°F. Thus even within the

The small grass Deschampsia antarctica *is one of the only two flowering plants found in the Antarctic. It may be a relatively recent arrival from southern South America, and now grows on several subantarctic islands and along parts of the west coast of the Antarctic Peninsula. The peninsular specimens are stunted compared with those from the islands and are rarely found in flower.*

Antarctic region climatic differences produce contrasts in the plant life.

South Georgia, lying between the northern limit of the pack ice and the Antarctic Convergence, is one of the warmest of the Antarctic islands. The peaks of its mountain backbone rear up to nearly 9000 feet and bear a permanent ice cap. Glaciers slide ponderously and imperceptibly down toward the sea, but during summer the lowlands are warm enough to thaw out. Meltwater is then abundant, and supplemented by heavy falls of wet snow, sleet, and rain. Together with a good, deep, organic soil, all this moisture gives rise to lush growths of tussac grass near the shores. A little above these meadows, where the drainage is somewhat better, there is a much more varied flora featuring some 20 species of flowering plants, together with ferns, mosses, lichens, and algae. Tussac grass, which usually grows to about eight feet but

South Georgia lies between the pack ice and the Antarctic Convergence, and is one of the warmest Antarctic islands. During the summer the lowlands are warm enough to thaw out. Water is abundant, and there is a lush growth of tall tussac grass (above). Southern fur seals (below) lie among the grass while they are molting, but until the introduction of reindeer from Norway (right) there were no land mammals on the island, except rats and mice that had arrived accidentally on ships.

sometimes reaches nearly twice that height, provides an excellent source of food for herbivores. With this in mind, whalers based in South Georgia introduced sheep, horses, upland geese, and mallard from the Falkland Islands. None of the descendants of these creatures has survived, but the whalers also brought reindeer from Norway, and the result was two herds of these herbivorous animals that now probably total several thousand individuals. Before the reindeer arrived, sealers evidently imported rabbits, but these grass-eaters survive only on an offshore islet.

East-southeast of South Georgia lie the South Sandwich Islands. Most are still volcanically active, and their ashy soils are cold and dry, but here and in the South Shetlands (some of which are also volcanic), steaming holes called *fumaroles* locally encourage a rich growth of mosses, although elsewhere these islands are mainly barren. The larger islands of the South Orkneys group are heavily glaciated, and pack ice surrounds the group for three quarters of the year. Here, though, Signy Island, in the lee of frigid Coronation Island, supports a lush growth of mosses and lichens that holds an immense quantity of water during the warmer months. On some parts of the island thick banks of moss peat bear witness to many centuries of plant production and decay.

So far little is known about the vegetation of remote Bouvet Island, almost 1250 miles east of the nearest of the South Sandwich Islands and some 1500 miles from the southern part of Africa. Even less is known of the botany of the Balleny Islands and Peter I Island, which lie respectively on and inside the Antarctic Circle. It is not even certain that they possess any plants at all.

In sheer mass the algae are the most abundant plants on the Antarctic continent itself, partly because they grow in many situations: on open

Algae can survive being frozen in the rare freshwater lakes of Antarctica. Here bottom-dwelling algae are seen through three feet of clear ice on Ross Island. Some of the lakes thaw during summer, but even those that do not may contain certain algae whose dark filaments can absorb enough radiation to be able to thaw the surrounding ice and allow them to live in a miniature aquarium.

ground, on snow, under ice, and in water. For example, blue-green algae appear as brittle rosettes, each neatly spaced from its neighbor, on sandy or stony ground where water is available from the thawing snout of a glacier or a snowbank. A bright green alga is found on damp rocks, and also flourishes beneath a layer of transparent ice built up from dripping meltwater. When conditions are just right, snow slopes may exhibit a beautiful, soft, rosy glow from the so-called snow algae growing on them. Apparently this somewhat unusual growth takes place only when the snow is softened by the sun to the point at which it begins to melt and yield many-branched streams. The snow algae seem to be restricted to coastal regions, for no one has reported seeing them on the Antarctic plateau. Red snow algae appear on mountain snows in many parts of the world, however, particularly in the European Alps.

Algae also grow in the freshwater ponds and lakes dotted, almost unbelievably, around the coastal areas of Antarctica. Many of these are permanently ice-covered, but some thaw out completely for a brief period at the height of summer. Even those that remain perpetually frozen often contain filaments of green algae embedded in the ice. Where these strands approach the surface their dark color absorbs enough solar radiation to thaw the adjacent ice, so that each floats in its own tiny aquarium.

Apart from the relatively richly endowed Antarctic Peninsula, the coastal areas of Antarctica possess only a few kinds of moss but many lichens. These plants grow most thickly near the nesting places of birds, especially penguins. On the other hand, substantial plants are unlikely to establish themselves actually in a penguin rookery, because these birds are very inquisitive and quickly pull up any plant that may have taken root. Moreover, as nesting proceeds, a rookery rapidly turns into an inches-deep morass of mud and droppings through which the penguins are constantly waddling. Thus, even where organic nutrients abound, Antarctic plants are often unable to use them.

The ground-level, single-layered vegetation of Antarctica is thus best described as a type of cold-desert tundra, formed of mosses and lichens, although liverworts and fungi also play a part. Few plants grow more than a few inches high. Nevertheless this is the major habitat of Antarctic invertebrates such as the tiny springtails, and even smaller mites, protozoans, rotifers, tardigrades (microscopic eight-legged arthropods), and threadworms. As an ecosystem this complex of plants and animals is simple, at least compared with the ecosystems of the multilayered vegetation of other regions, with their numerous and varied kinds of animals. Antarctica's largest native land animal is a wingless fly about one fifth of an inch long, and the largest carnivorous land animal is a mite about one thirtieth of an inch long that feeds on herbivorous insects. Antarctica's permanently frozen ponds normally lack animal life, but ponds that do thaw out frequently contain a flourishing population of minute invertebrates including bacteria, protozoans, rotifers, tardigrades, threadworms, and flatworms. Freshwater pools on the Antarctic Peninsula sometimes even contain a fairy shrimp about three sixteenths of an inch in length, the only freshwater Antarctic crustacean known.

Terrestrial vertebrates are completely absent from Antarctica, whereas most kinds have been recorded from the Arctic, where only the amphibians and reptiles—the so-called cold-blooded animals—are almost entirely lacking. (Even here the wood frog lives as far north as Alaska and Labrador, and the adder ranges north to Lapland.) As later chapters show us, the richness of Antarctica's wildlife is in the surrounding seas; that of the Arctic lies at least partly on the land.

A fairy shrimp is the only Antarctic freshwater crustacean known. It lives in seasonally thawing pools on the Antarctic Peninsula, together with various other tiny aquatic animals.

Land Mammals

Compared to more hospitable parts of the earth, the polar regions possess conspicuously few species of terrestrial mammals. Indeed, Antarctica has no native land mammals at all. This is partly because the climate is so harsh and unrelenting that they would have a hard time surviving, particularly in the winter. They could not even migrate to and from warmer continents as visitors. At least 700 miles of rough sea—an effective barrier for any land mammal without a boat—separates Antarctica from South America, and the distance from the southern tip of Africa is more than three times as great. In any case, too little vegetation grows in the Antarctic to feed large numbers of mammalian herbivores and, without these for prey, few land-bound carnivores could exist. True, cold-adapted predators such as polar bears and wolves could subsist on emperor penguins and seals during the winter and on nesting seabirds in the summer. But, because Antarctica is surrounded by oceans, terrestrial carnivores cannot reach it without the aid of man. Indeed, it is unlikely that Antarctica has ever possessed any native land mammals. Small dinosaurs and mammal-like reptiles were certainly present when climatic conditions were milder, but that was some 200 million years ago.

The Antarctic Treaty of 1961 expressly forbids the deliberate introduction of animals and plants to Antarctica because of a very real threat of upsetting the delicate natural balance of the region. Even the huskies kept at various research stations pose a problem, for the possibility of some escaping and forming a "wolf-pack," to survive and breed as long as a food supply of penguins lasts, is not as remote as it may seem. In 1958, for example, two out of 15 sledge dogs, abandoned because of bad weather by Japanese leaving their station for the winter, survived to greet the returning scientists the following spring. In 1960, Sir Vivian Fuchs lost one of his sledge dogs on the Antarctic Peninsula. It turned up at the ship 60 days later, fat and healthy, having traveled across 150 miles of ice. Presumably it had fed on seabirds or seals along the route.

Land mammals apart from the dogs *have* been introduced, but only to certain subantarctic islands. Some of these introductions were accidental. For example, rats and mice escaping from whaling and sealing ships invaded and colonized many islands, including South Georgia and the Kerguelen Islands. They quickly became, and still are, predators of petrels, which nest in burrows and are highly vulnerable. Because there are no predators to control these rodents, their numbers have grown very large, presenting a serious threat to bird populations. The introduction of domestic cats to keep down rats and mice proved even more disastrous, for the cats also developed a distinct appetite for petrels.

Deliberate introductions have been made, however, a number of which have been successful from a survival point of view, although in some instances they have threatened plant communities through overgrazing. A small group of domestic sheep was introduced to the Kerguelen Islands in 1948, and had grown to a population of nearly 800 animals by 1970. Reindeer introduced to South Georgia and the Kerguelen Islands have also done well.

By contrast, the north polar regions possess a moderate number of land mammals of their own. There have been, however, a few reintroductions of species such as the reindeer and the musk-ox, now reestablished in areas that had lost them through natural causes or overhunting.

Conditions for life are more tolerable in the Arctic than in the Antarctic. The climate in the north is less severe than in the south. During the short northern summer, vegetation is plentiful enough to feed large numbers of herbivores, which in their turn become the food of carnivores. Furthermore, the continents having an Arctic fringe are continuous with milder regions further south to which many of these animals can migrate to avoid the harsh winter conditions.

Even so, life on the tundra is not easy. Perhaps the greatest problem facing northern tundra species is not so much the winter cold as the great seasonal fluctuations in temperature. Temperatures in the Arctic may swing from 60°F in the summer to −70°F in the winter—a difference of 130°F. There are few species of land mammals that have adapted to withstanding such cold in winter and yet are able to avoid overheating in the summer—a fact that helps to explain why the

The polar bear, an inhabitant of the tundra and pack ice of the Arctic, is the world's largest land carnivore. Adult males may weigh up to 1500 pounds. Although they are somewhat solitary by nature, polar bears will often play with others of their kind.

richness of Arctic animal life lies not in a wealth of species but in many individuals of a few forms.

Of all the northern species of terrestrial mammals, perhaps the one best adapted to a year-round life on the tundra is the shaggy-coated musk-ox. It is a relic of the past; together with such mighty creatures as the mastodon and the mammoth it once roamed the plains of Europe, Asia, and North America, following the glaciers as they retreated northward at the end of the last ice age. It then died out in Eurasia, its extinction perhaps accelerated by Stone Age man, and by the middle 1800s it survived only in Greenland and northern Canada. Since then, musk-oxen have been reintroduced to many areas, including Iceland, Spitsbergen, Siberia, and Alaska.

The musk-ox looks somewhat like a buffalo gone wrong, but in fact is more closely related to sheep and goats than to cattle. Its marvelously warm coat, perhaps the thickest and longest of any animal, is composed of two layers: an inner layer of very dense, fine, soft, light-brown fleece, covered by an outer one of coarse, blackish-brown guard hairs, which are about six inches long on the back but may be as long as two or even three feet on the chest, neck, and hindquarters. With such a coat the musk-ox seems impervious to cold, standing out on the wind-lashed tundra during the darkest and coldest days of the winter when other animals have sought the shelter of valleys or burrowed in the snow. But the coat that traps vital heat in winter may make the musk-ox uncomfortably hot in summer, although it starts to molt in April and

Male musk-oxen defend their females and young from attack by predators by forming an outward-facing ring around them (left). Unfortunately, this makes them easy targets for human hunters. Musk-oxen calves look very much like miniature versions of their parents (below). They shelter under their mothers' coats during storms.

the new undercoat does not appear until mid-July.

Unfortunately for the musk-ox, man also found its coat exceptionally warm. Adults of whole herds were shot; their fleecy skins were turned into sledge blankets, their meat sustained Arctic explorers and their dogs, and the calves were sent to zoological gardens, where they seldom survived for long. The originally sparse population fell alarmingly and by 1930 it was estimated that only 500 animals remained on the mainland of Arctic Canada. Since 1947, however, Canada has enforced absolute protection, and the species is making a slow but sure comeback. In 1974 estimates placed the mainland population at about 1500 animals, and a further 8500 were believed to live on Canada's Arctic islands.

Although musk-oxen are fairly large animals—the bulls averaging five feet at the shoulder and the cows four feet—they have relatively short, stubby legs. This is an advantage in a cold climate, for heat loss is reduced by a smaller surface area. The hair-covered tail is also stubby (no more than four to six inches long), and the ears project only slightly from the coat. The hooves are broad and sharp-rimmed and provide good traction on rock and ice.

Musk-oxen are gregarious animals, living in herds that may contain from three to 100 individuals but generally average around 15. When danger threatens, the bulls form an outward-facing ring around the calves—an effective defensive posture against their natural enemy, the Arctic wolf, but absolutely useless against men with firearms. Adults of both sexes shelter the calves during most storms, but in particularly bad weather the males bear the brunt of nature's fury, shielding the rest of the group with their bodies, standing there for days if necessary.

During the rutting season in early fall, tremendous clashes take place between bulls challenging one another for possession of a herd. Two opponents stand face-to-face and charge, their heads meeting with a resounding crack. The rivals then back off and perhaps nibble blades of grass. Suddenly they whirl and charge once more. This continues until the weaker of the two turns and flees.

Calves, one to a cow, are born in April or May after a gestation period of eight to nine months. At birth the calf lacks the long guard hairs of the adult, and these do not start to grow until the beginning of the second winter; but when it has dried after birth, its thick, short natal coat of curly, dark-brown hair protects it well from the cold. If conditions become too severe, however, the calf seeks shelter under its mother's long winter coat. Musk-oxen herds are often on the move so it is a distinct advantage that a calf is able to follow its mother only an hour after birth. It grows rapidly on her rich milk and within a week or two is nibbling grass.

These animals migrate only in a limited sense. During the summer they dine richly off the grasses, sedges, and willows in river valleys and along lake shores, building up food reserves within their bodies. In winter they remain on the tundra, but shift to hilltops and slopes where the winds free the vegetation of snow. Even so, winter is a lean time and by early summer their fat reserves have been depleted.

By contrast, the North American caribou and Eurasian reindeer (one species with two names) undertake regular southerly migrations of up to 800 miles in the fall, returning north to the tundra the following spring. This species is unique among deer in that both sexes possess antlers, although those of the female are much smaller than those of the male and are often no more than forked spikes. In color, reindeer vary from white to nearly black, but are usually brownish or grayish with lighter underparts in summer, becoming lighter overall in winter. They are highly specialized for Arctic conditions, with a heavy coat of woolly underfur protected by straight, stiff guard hairs. The hairiness of the muzzle and the shortness of the ears and tail give added protection against cold. To reduce heat loss from their relatively long legs the temperature of these appendages is kept as much as 50°F lower than that of the body—an adaptation that is shared by many Arctic animals including wolves and birds. The broad, flat hooves are deeply cleft so that they spread out, giving the animal maximum support on snowy and swampy ground.

During the summer caribou eat grasses, aquatic plants, berries, twigs, and lichens. Mating takes place in September or October, and the males spend most of this time making a great fuss: rushing around, battling with other males, panting, and bellowing, and generally spending little time feeding. Caribou are polygamous and the bulls mate with from five to 40 cows.

Reindeer undertake mass migrations (above), and in 1969 traffic on a North Siberian railroad track was stopped for six nights to allow them to pass. A very large proportion of the estimated 300,000 animals inhabiting the Taimyr Peninsula migrated in 1968, and they are seen here on their way back to their northern breeding grounds. The Lapps have domesticated this species for many hundreds of years, but have never been able to breed out the strong migratory instinct. They follow their herds on the long, cold treks, stopping to rest only when the reindeer do. A close relative, found on high Arctic islands, is the smaller nonmigratory Peary's caribou (left), named for the American explorer Robert E. Peary who was the first man to reach the North Pole.

Once the breeding season is over, herds gather together in large numbers for the fall migration toward the wooded areas farther south. These herds, led usually by the males, travel to their winter feeding grounds along definite routes, crossing lakes and rivers in their path and covering up to 100 miles a day. On arrival, they feed almost exclusively on lichens, from which they must clear the blanketing snow.

In North America the annual southerly trek was once extremely important for the welfare of the Eskimos and northern Indians, who hunted the migrating caribou for winter food and clothing. While the numbers killed during these hunts depended solely upon the bow and arrow, and the skill of the archer, the centuries-old balance between predator and prey was preserved. After the arrival of firearms with the Europeans, however, the scales dipped alarmingly in favor of the predator and the numbers of caribou were reduced so drastically that many of the old migration routes were forsaken. On mainland Canada the caribou population slumped from an estimated 3 million to no more than 277,000 by 1955. Happily, a survey taken in 1967 indicated a population of 387,000 animals, so it seems that numbers are rising again.

Toward summer, caribou herds gather together once more and head back to the tundra. This time, however, they are led by pregnant females seemingly driven by the desire to reach their calving grounds in the north. The calves, one or two to a cow, are born between mid-May and the end of June, after a gestation period of seven and a half to eight months. They are very precocious, and are able to walk a few hours after birth, to outrun a man after 24 hours, and to keep up with a running herd within a few days. Suckling continues for about two months, but a youngster may supplement its mother's milk by nibbling at a few plants after a couple of weeks.

Apart from man, the caribou's main enemy is the wolf, a well-known, much-misunderstood carnivore resembling a large, pale, German shepherd dog. Some wolves live on the Arctic coasts of Alaska, western Canada, Europe, and Asia. In Arctic Europe, however, they are heading for extinction: for instance, by 1963 only about 25 survived in Sweden and most of those were in the extreme north. The situation is better in North America, where wolves are now protected, but even so their numbers have declined rapidly over the past century or so.

Above: the wolf has been the subject of horror stories and fairy tales down through the ages, but in fact it is a shy family-loving animal. Only where man has deprived them of their natural prey have wolves turned to killing domestic animals.

Wolves live in family units called packs, which consist primarily of parents, their offspring, and near relatives such as aunts, uncles, and grandparents. In each pack the strongest and largest male is usually the leader, followed by the younger and older males, the leader's mate, the remaining females, and finally, in order of strength, the pups. A strong bond of affection exists between the pack members and there is much romping and general playfulness. But the adults also show a great deal of respect toward one another, and all are responsible for protecting, feeding, and educating the young. In fact, wolves display many of the characteristics that man so much admires in his own species, and it has been suggested that we may learn more about man's own evolution and development by studying this carnivore than by studying the lives of the primarily vegetarian great apes.

Yet to many people the wolf is still the arch-villain of the fairy tale—a ruthless killer that attacks man without provocation and that is solely to blame for the rapid drop in populations of herbivores such as the reindeer. Bounties have been placed on its head, tens of thousands have been poisoned, and many thousands more shot. Fortunately for the wolf, some of the myths are now being exploded. All who have studied its habits in the wild agree that it is remarkably shy of humans and that, in every case reported in North America, attacks on man have been made by diseased (rabid) animals. The eerie howl that has terrified man for centuries is not some "devil's war cry," but an important communication link between pack members, who often become separated while hunting. Wolves hunt only for food and may perform a valuable service to their prey species by culling the weak and sick animals. Where wolves have been exterminated their prey also may be in danger of extinction. The unchecked populations of prey animals may build up until they destroy their own habitat; lacking food, their populations then crash.

Man is also responsible for the rapid decline in the numbers of another Arctic and subarctic carnivore, the wolverine. This solitary animal is found on the southern tundra and in the northern coniferous forests of Alaska, Canada, and Eurasia. Its valuable, shaggy, dark-brown fur has the peculiar property of retaining less condensation than any other fur and is used to edge parka hoods. The wolverine is also hunted because it destroys many creatures prized by trappers, for this short-legged, thick-bodied creature less than three feet long is said to kill more than any other predator—a fact that explains its alternative name of glutton. Its feet are equipped with strong claws and, although few individuals stand more than 14 inches high at the shoulder, this extremely powerful animal will attack almost anything, including bears. It is not, however, a fussy eater, taking (among other things) carrion, eggs of ground-nesting birds, wasp larvae, and berries, and playing in many ways the same role in the north that the hyena plays in the tropics. The wolverine can climb well and, like the leopard, has been known to cache its kill in a tree, but it lives mainly on

Below: the wolverine is a relatively small but extremely powerful animal that is incorrectly said to kill for the sake of killing, but it certainly does steal from traps when possible. It has been hunted by man to the verge of extinction for its valuable fur.

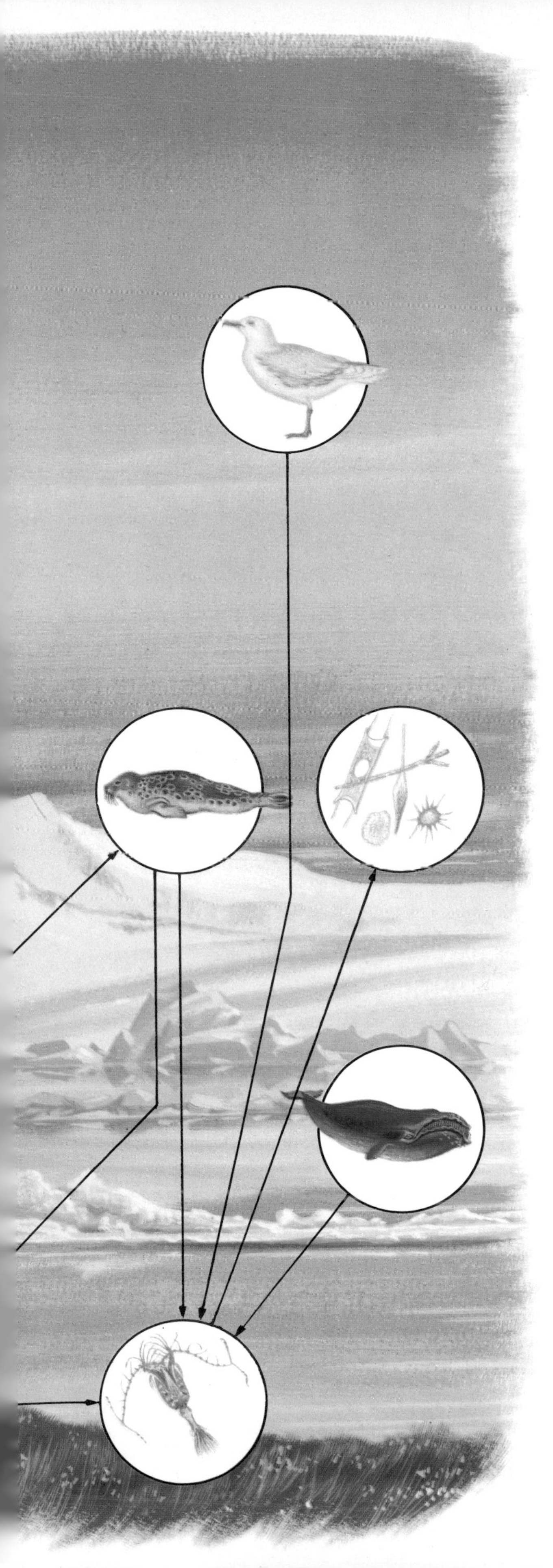

An Arctic Food Web

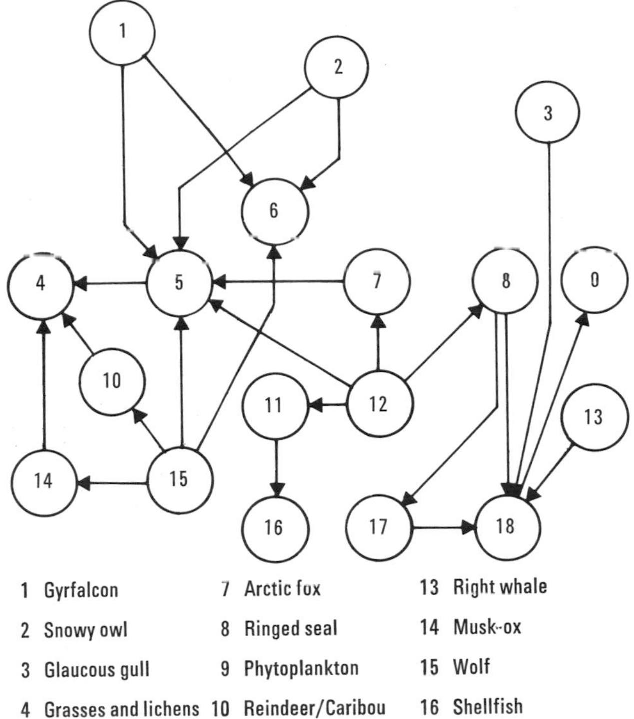

1 Gyrfalcon
2 Snowy owl
3 Glaucous gull
4 Grasses and lichens
5 Lemming
6 Ptarmigan
7 Arctic fox
8 Ringed seal
9 Phytoplankton
10 Reindeer/Caribou
11 Walrus
12 Polar bear
13 Right whale
14 Musk-ox
15 Wolf
16 Shellfish
17 Polar cod
18 Krill

This Arctic food web has sea and land aspects. At sea, phytoplankton and other plants feed krill and shellfish and, indirectly, fish, seals, whales, and gulls. On land, grasses, lichens, and other plants support musk-oxen, caribou, lemmings, ptarmigan, Arctic foxes, wolves, gyrfalcons, and snowy owls.

the ground, where it moves at a loping gallop.

Two of the world's largest land carnivores are found in the Arctic: the brown bear and the polar bear. The brown bear occurs throughout most of Eurasia, extending into the north polar regions over much of its range, and one form—the Barren Grounds grizzly bear—inhabits North America's Rocky Mountain region, Alaska, and the western Canadian Arctic. The few accurate measurements that exist for this large carnivore indicate that it grows to a length of about eight to 10 feet. Comparatively little is known of its habits, but it appears to be active mainly in the evening, at night, and in the early morning, spending much of the day sleeping. Around mid-November it retires to a den and spends the winter dozing, rarely emerging before April.

The polar bear occurs farther north than the brown bear and inhabits all Arctic seas and

The Alaskan grizzly bear is among the largest of the brown bears. It sometimes eats carrion, as it is doing in this photograph, but it will feed on almost anything edible, including such varied items as berries and salmon. Its habits are still relatively little known to naturalists, but it appears to be active mainly between dusk and dawn. It spends the long northern winter dozing in its den.

coastlines. Males usually grow to a length of about nine feet, but immense specimens up to 15 feet long have been recorded. The yellowish-white coat of long, coarse, oily, waterproof fur is considerably thicker in winter than in summer and, together with a blubber layer beneath the skin, insulates the animal well against the cold. Although providing effective camouflage against a background of snow and ice, its creamy color makes the animal conspicuous on snow-free summer tundra. Much of the sole of each large foot is thickly haired, providing both insulation and a secure grip on the ice. To protect the eyes from possible sunblindness the eyelids apparently act as sunglasses.

Despite their somewhat lumbering appearance, polar bears are surprisingly active animals and can outrun a caribou over a short distance. They are very strong, though relatively slow, swimmers, using a dogpaddle with the forelimbs and trailing the hindlimbs; they can float for hours. Polar bears have a well-developed sense of smell and are said to be able to detect their prey from 20 miles away. Their diet is variable and includes young walruses, caribou, musk-oxen, Arctic hares, nesting birds, fish, carrion, berries, and seaweed. But perhaps their favorite food, particularly in winter when it is easier to find, is the

Polar bears are great wanderers and often travel miles across the ice in search of seals or fish for food (above). They are equally at home in the water (below), and are protected from the icy cold of the Arctic regions by an oily, waterproof fur that is able to trap an insulating layer of air next to the creatures' skins.

ringed seal. A bear will sit patiently by a seal's breathing-hole for hours, waiting for an unlucky victim to emerge. As soon as a seal's head appears above the surface, the bear will grab it with one swipe of its huge paw and haul the animal bodily through the hole, often exerting enough force to break the skull.

Polar bears are relatively solitary animals, and males and females spend most of the year apart except for a short two-week period during the breeding season in April or May. In late autumn pregnant females excavate dens where they spend most of the winter dozing, living off fat reserves built up in their bodies during the summer months. There, between mid-November and early January, they give birth to from one to four blind and deaf cubs, each weighing less than two pounds. Mother and cubs remain in the den until late March. By then the young have increased their weight nine times.

Unmated females and young males often spend much of the winter under cover, especially if the weather is particularly bad, but adult males seek shelter only during the worst storms. In the winter they range widely in search of food and have been recorded as far inland as 75 miles—an unusual occurrence for an animal that appears to prefer the edge of the Arctic pack ice.

Trotting behind these wandering polar bears, keeping a respectful distance, will sometimes be seen the little, white Arctic fox, hoping for leftovers from a kill. This charming creature lives in Greenland, Iceland, and Spitsbergen, and on the tundra of Europe, Asia, and North America. The summer coat is grayish-yellow but in winter there are two color phases: some foxes have a

blue-black to pearl-gray coat, whereas others become a beautiful, almost pure, white. The blue-phase foxes seem to favor coastal areas and the Icelandic population consists entirely of this type, but the white-phase foxes predominate in central Eurasian and North American Arctic. In both cases the winter coat is very long, providing the animals with good insulation. The soles of the feet are also covered with fur and provide a good grip on snow and ice. To reduce heat loss the muzzle and well-furred, somewhat rounded, ears of this species are shorter than those of all other species of fox—a good example of Allen's Rule, which states that the extremities, such as legs, ears, and noses, of animals living in cold climates are shorter than those of their relatives from warmer regions.

Arctic foxes store considerable quantities of food under rocks and in crevices for the winter. Their diet is variable. They are the main predator of many species of birds and also eat hares (of which there are three Arctic species), fish, reindeer calves, berries, shellfish, and carrion. When they can, however, they live chiefly on lemmings, catching them on the open tundra in summer, and digging them out of their tunnels and nests under the snow in winter—a chore that the foxes tackle with feverish activity, furiously scraping away with their forepaws.

The chunky, stubby-tailed little lemmings are preyed upon by a wide variety of polar carnivores besides the Arctic fox. These predators include the mouse-sized least weasel (the world's smallest carnivorous mammal), the ermine or stoat, the wolf, the wolverine, and the brown bear. In addition lemmings are included in the diet of birds

Both the Arctic hare (above) and the Norwegian lemming (below) are among the creatures eaten by the carnivorous Arctic fox (left). Several polar birds and mammals, including the fox and the hare, have two different color phases for concealment: a dark one during the summer months and a white one in the winter. Otherwise, such creatures would be dangerously conspicuous during one half of the year.

Right: the Arctic ground squirrel is the only Arctic mammal that truly hibernates. Toward the onset of the severe northern winter weather it builds a special den in the ground, curls itself up in a ball, and sleeps soundly until the following spring.

such as the snowy owl, gyrfalcon, and skuas.

Four species of lemming live in the countries bordering the Arctic, and the periodic peaks and troughs in the numbers of these hamster-sized rodents may well be the key to the ecology of the tundra. Lemmings can multiply so rapidly over three to four years that they eat all the available roots, stems, and leaves that make up their food. When this happens, the majority of the starving lemmings abandon their homes and migrate to found new colonies in vegetation-rich areas. Their routes may be blocked by the sea, lakes, or rivers. Nevertheless, driven on by instinct, they attempt to cross the water and thousands are drowned—there is no truth in the tales of mass suicide. During the years of high lemming density, predators move deep into the tundra, lured from the south by an abundance of prey. These predators produce extra-large litters during this time, and plentiful food means that most of their

The least weasel (above) is a formidable predator despite its small size. It is the world's smallest carnivorous mammal. A close relative, the ermine (right), reveals its presence here by its black eyes and nose. Its pure white coat (with black tail-tip), so attractive to humans, blends effectively with the snow.

young survive. Suddenly the lemmings have gone. A year later predator populations slump too, for food is now scarce, especially during the winter. Many die of starvation and mothers may kill the young they can no longer feed.

A fifth species of lemming that lives in the north does not show this kind of "boom and bust" population fluctuation. This is the collared, or snow, lemming, the only rodent that turns white in winter. It lives throughout Arctic Canada and the Arctic USSR, along the northern coast of Greenland, and on other Arctic islands.

Unlike larger Arctic animals, lemmings are too small to protect themselves from the winter cold merely by growing a thicker coat or developing a thicker layer of blubber; nor do they hibernate. Instead, they make extensive burrows through the soil or runways among the vegetation under the snow, which help to protect them from heat loss through wind and from predators.

Perhaps surprisingly, there is only one Arctic species that truly hibernates—the Arctic ground squirrel or *siksik*, as the Eskimos call it. This noisy rodent is extremely active all summer, but in September or October it builds a special den off its main burrow, some 24 to 30 inches below the surface, lines it completely with grasses, reindeer hair, leaves, and lichen, curls up into a ball with its tail over its head and shoulders, and goes into a long, deep sleep. It remains this way for some seven months of the year and during this time its body temperature drops from 97.5°F to 62.6°F, it takes only two or three breaths a minute instead of 60, and its heart beats only once every two or three minutes.

It is not known what triggers their reawakening but in late April or early May, before the snow has completely disappeared, these little animals burrow their way up to the surface. Spring has come again to the Arctic.

Birds

As soon as the long winter night is over, and when enough snow has melted to reveal grasses, mosses, and lichens, hordes of birds arrive in the Arctic. They fly in from central Asia and North and Central America, and even from as far away as the Antarctic. On the Arctic tundra, coasts, and islands, these summer visitors join the ravens and snowy owls that have hardily overwintered in darkness amid the snow and ice. Nesting begins at once, and immediately the young have hatched activity reaches its climax. The breeding season is desperately short, but long hours of summer daylight speed up the growth of plants and insects, and the birds' eggs hatch when such sources of food are most plentiful. Lakes, streams, marshes, and peat-bogs are among the region's richest natural larders, teeming with insects and other small animals. Consequently, these summer wetlands attract ducks, geese, swans, waders, and divers. Relatively fewer species feed on land, but wagtails, pipits, bluethroats, thrushes, Arctic warblers, and finches are common. On the rocky coasts and cliffs almost every ledge, cranny, and grassy slope forms a home for squabbling groups of gregarious seabirds such as cormorants, guillemots, razor-billed auks (razorbills), dovekies (little auks), northern fulmars, gannets, puffins, and gulls. Here and there, sandy shores support colonies of bickering terns that scream their distrust of intruders.

While this hectic race against time is going on in the far north, at the other end of the world it is winter. The Antarctic continent is locked in the grip of ice both on land and by sea. Unbroken darkness covers most of the region for weeks on end, except for starlight and the flickering, shimmering aurora australis when the sky is clear. The cold is bitter even without the wind that is usually blowing. Yet here, in the Antarctic winter, birds are nesting—if you can call it nesting when the site is a shelf of ice, and the nest is a parent bird's feet. This coldest, darkest, midwinter period is the emperor penguin's breeding season, timed so that the slow-developing young can be launched on the world when their food supply is best assured. Upon the arrival of spring, other seabirds pause in their travels

Pink-footed geese flying to their feeding grounds at dawn. Geese are grazing social waterfowl. A number of species breed on the northern tundra, migrating southward along well-known flyways to avoid the winter. Their northward flights often follow different routes, but the breeding grounds, wintering areas, and flyways of some species have been changed by human interference.

on or over the oceans, and then engage in the frantic rush to find or fight over nest sites and mates, lay eggs, hatch their young, and raise them so as to be away again before another cruel winter sets in.

This summer migration into Antarctica broadly parallels the events we described in the Arctic. But, in fact, the bird life of the two polar regions shows more differences than might be expected. This is because the regions themselves differ. It is true that in both places much of the land is barren, and for most of the year the adjacent seas are frozen. But one region is water surrounded by land and the other is land surrounded by water, and this contrast powerfully affects their bird populations.

There are three main contrasts between Arctic and Antarctic bird populations. First, the Arctic has far more species present for at least part of the year; secondly, it possesses many more species of land birds than the Antarctic; and thirdly, Antarctica has many flightless species, whereas the Arctic regions have none. (The great auk, the north's sole flightless bird, was wiped out by man more than a century ago.)

It is not too hard to see how these differences have come about. Let us first take numbers of breeding species. Birds nest only where there are

suitable sites and where they can get enough food for their chicks. In the north there are plenty of nesting sites, both on level or gently sloping ground and on the cliffs of the mainland and islands. In the Arctic summer, food is reasonably abundant on land, and plentiful in the freshwater lakes, rivers, and marshes, and in the neighboring ocean. Largely because of these reasons the Arctic possesses as many as 120 species of breeding birds—a poor total compared with the 1700 that occur in Colombia, the world's richest region for birds, but about seven times more than Antarctica's tally. Few of these Arctic nesters are equipped to overwinter in the far north. In the fall, most migrate south to temperate or tropical lands. Even rather weak fliers can survive the journey, for most Arctic migration routes involve no hazardous flights across oceans. Instead, many flyways lie almost entirely over the land, which means that land birds can rest and feed on the way.

On the Antarctic continent breeding conditions are much less favorable. Potential nesting sites free of ice or snow are rare on the mainland and they do not abound on the islands. Even in summer, sources of food on the land are limited. Furthermore, the nearest continents are hundreds, or even thousands, of miles away to the

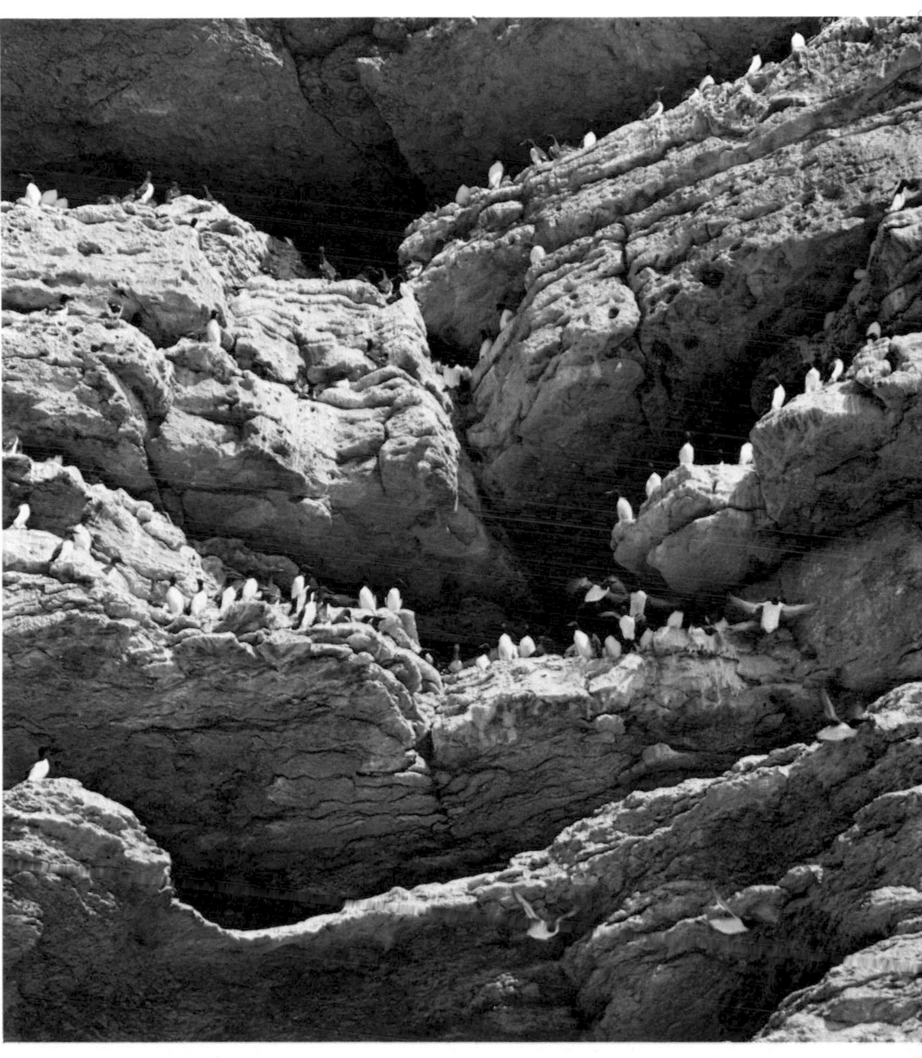

The Canadian tundra is an enormous patchwork of lakes and rivers (left), created often as a result of poor drainage due to the underlying permafrost. However, these wetlands offer a rich food supply, especially of water-breeding insects, and consequently are the summer habitat of ducks, geese, swans, waders, and divers. The offshore waters of the Arctic Ocean are also highly productive of food, and during summer the cliffs are the breeding grounds of murres (above) and other seabirds.

The lower levels of the Antarctic island of South Georgia thaw out in the summer (left), and in some parts there is a rich vegetation. In these areas live four species of land birds, including the South Georgia pipit (above) and the South Georgia pintail (below). These are the southernmost-dwelling representatives of the songbirds and the ducks respectively.

north over gale-swept seas, so that migration to and from Antarctica is difficult. Not surprisingly, then, the south polar regions as a whole possess fewer than 50 breeding species, and only 16 hardy kinds nest on Antarctica itself.

The harshness and remoteness mean that land birds are especially uncommon in the Antarctic. Indeed, the area has only one species of native songbird, the South Georgia pipit, which, as its name suggests, is restricted to the island of South Georgia. There used to be two land birds, a rail and a ground parakeet, both inhabiting Macquarie Island, 850 miles southeast of Tasmania. But both were exterminated about 1890 by sealers and the cats and rats that they introduced to the island. Shore birds are represented by two closely related species of white, superficially pigeonlike birds called sheathbills, sole members of their family, which, incidentally, is the only family peculiar to the region. Finally, in contrast to the Arctic's rich variety of ducks and geese, the Antarctic has only three species of ducks. The South Georgia pintail is closely related to the Chilean pintail, and probably arrived fairly recently as a vagrant on the west wind from South America. The Kerguelen pintail, on the other hand, probably derives from the Northern Hemisphere, because it appears to be most nearly related to the northern pintail, a bird that breeds up to 80° North in Canada.

All the remaining birds of the Antarctic are seabirds, deriving most or all of their food from the oceans and coming ashore mainly to breed. These birds include albatrosses, fulmars, petrels, prions (small gray birds), cormorants, skuas, a gull, terns, and, of course, penguins. But although the number of Antarctic species is small, the total of individuals is vast. For instance, Wilson's petrel, one of the tiniest of all seabirds, breeds so abundantly in Antarctic regions that it may be the world's most numerous seabird. Indeed, apart from the domestic fowl, with possibly about three thousand million individuals, Wilson's petrel may even be the commonest of all birds, although it has several strong rivals for that title, including the common starling and the house sparrow.

Antarctica's most remarkable birds are undoubtedly those unable to fly. All flightless birds are probably descended from ancestors that *could* fly: for example, the great auk was closely related to the murre and razor-billed auk, and Antarctica's penguins themselves derive from flying birds that also gave rise to the albatrosses and fulmars. We do not know exactly why some birds lost the power of flight, but we do know that most of the world's flightless birds live or lived (for many are extinct) on isolated oceanic islands. It may be that their forebears arrived by flying to these islands, and that these birds' descendants gradually lost their ability to fly. Because there were few, if any, mammal predators and competitors, and because the birds fed on the ground or in the water, flight was no help in escaping enemies or seeking food. Moreover, flying birds living on windswept islands face a greater risk of being blown away than birds confined to land or water. These conditions broadly apply to Antarctica, where penguins reach their greatest abundance. In the Arctic, on the other hand, mammal predators such as the polar bear, wolf, and Arctic fox would place most kinds of flightless bird at a disadvantage. All this helps to explain why Antarctica has many flightless birds but the Arctic has none.

We have sketched the broad contrasts between the bird populations of the Arctic and Antarctic regions. Now let us take a closer look at the birds of each region in turn. For simplicity, we may divide Arctic birds into four categories: land, freshwater aquatic, shore, and sea. (These groupings do not always hold good. For instance, we shall see that some birds ranked as freshwater aquatic spend much time in salt water.)

Arctic land birds include what is probably the world's northernmost species of songbird, the snow bunting or snowflake. The high, "dancing" flight of flocks of these small black-and-white birds makes them appear at a distance like drifting snow. Snow buntings have a beautiful, rapid, high-pitched, almost larklike song that is sung on a downward glide following a circling display flight. They live all around the Arctic, and are somewhat less common in the subarctic, their distribution being largely dictated by temperature. In July they are restricted to an area south of the 35°F isotherm and north of the 59°F isotherm. But some venture very far north. A Soviet drifting-floe station observed snow buntings quite close to the North Pole, and, in the spring of 1954, observers pinpointed them at 86° 59′ North, just over 3 degrees south of the Pole. Rough, stony areas, mountain slopes, and cliffs provide these small birds with the crevices they need for nesting, and minor hillocks serve as singing stations. In some ways, the snow bunting fills in the Arctic the role that its relative the house sparrow plays in more temperate lands, in that it fearlessly builds its home close to man, invading camps and villages to nest in man-made structures. All other Arctic birds nest on, or very close to, the ground, for of course, the Arctic is treeless.

Arctic land birds also include two species of game birds: the willow ptarmigan or willow grouse, and the rock ptarmigan. The willow ptarmigan is widely distributed in northern parts

The snow bunting or snowflake (above) is the most northerly known species of songbird, and it has been seen quite close to the North Pole. In the winter it migrates southward, usually in large flocks, and eats the seeds of any plants that are standing out above the surface of the snow.

Wilson's petrel or Mother Carey's chicken (below) feeds while apparently "dancing" over the water, picking up planktonic organisms as it goes. This tiny bird breeds on the Antarctic continent, and on almost every island in the region.

of North America, Europe, and Asia. It remains in the Arctic throughout the winter, but during cycles of abundance may migrate far south of the northern timberline. The willow ptarmigan prefers low, moist, and better vegetated tundra than the rock ptarmigan, and especially favors areas where the dwarf willow grows, for the twigs, buds, and leaves of this plant form its main food throughout the year. The range of the rock ptarmigan extends farther north, and those birds living as far south as British Columbia, Spain, central Asia, and northern Japan occur only on high mountain slopes. Both species are mainly white in the winter, and largely brown or gray in summer, with an intermediate plumage in the spring and fall. Indeed, in spring, summer, and fall they appear to be in continual molt. These seemingly eccentric seasonal changes of color have great camouflage value, for white blends with the snow in winter, and brown or gray merges with the tundra vegetation at other times of the year. In the high Arctic, where there is always some snow, the rock ptarmigan remains white at all seasons. Such camouflage is particularly valuable for the ptarmigans because they are poor fliers, especially vulnerable to birds of prey such as the gyrfalcon and the snowy owl, and to carnivorous mammals including the wolf and the Arctic fox. In winter, ptarmigans are protected from sub-zero temperatures by their extremely dense plumage, even their legs and feet being well feathered. (Feathered feet, incidentally, also serve them as snowshoes.)

The sole Arctic representative of the crow family is the raven, the largest of perching birds with a length of just over two feet. Omnivorous scavengers, ravens wander over all kinds of terrain, probing inland and scouring the coast. They are adaptable creatures that have become numerous near settlements, where they raid rubbish dumps for food. Pairs also often establish themselves near seabird colonies, which they plunder. The raven's ability to survive on anything edible makes it one of the few kinds of birds able to overwinter in the Arctic, and ravens are probably the first to nest there in spring.

Other perching birds that inhabit or visit the Arctic include the wheatear, which occurs in inland and coastal regions, particularly among piles of weathered rock; the water pipit, widely distributed in the Arctic of both the New World and the Old, and closely related to Antarctica's South Georgia pipit; and two species of redpoll.

The hoary redpoll (also known as the Arctic or Hornemann's redpoll) actually overwinters in the high Arctic. It is surprising that a bird weighing only one third of an ounce can survive a Greenland winter. In general, overwintering birds and mammals tend to be large—for example, the rock ptarmigan weighs somewhat more than one pound and the snowy owl and raven each weigh more than three pounds. This is because a big warm-blooded animal can maintain its body temperature more easily than a small one. In any creature, heat is lost through the body surface, and the surface of a small creature is proportionally much greater than that of a large one. Warm-blooded *(homeothermic)* forms, (birds and mammals) have evolved complex means of at least broadly maintaining the temperature of the body core: that is, the deeper tissues below the surface of the skin. The generally rapid rate at which warm-blooded animals' life processes proceed produces a lot of heat, but it also calls for a relatively large amount of food, and this must be efficiently digested. Hoary redpolls feed primarily on willow and alder seeds, which contain highly nutritious food reserves laid down by the parent plants. Moreover the seeds appear to be easily digested by the birds. Naturally, the more food a redpoll can find and digest during the brief periods of winter daylight the better, and it is a remarkable fact that the redpoll's digestive system seems to become more efficient during winter. The heat produced by the resulting high level of chemical activity inside the bird's body is largely prevented from leaking out into the air by a layer of feathers. This proves such an effective insulator that it causes overheating if

The raven is the sole Arctic representative of the crow family, and lives in mountainous country and along the sea coasts. In the Arctic, it often chooses to nest on cliffs that are frequented by seabirds whose nests it raids. Ravens are omnivorous birds and overwinter in high latitudes. They are probably the first birds to make their nests in the north polar region in the spring.

the birds are exposed to very high temperatures.

Two birds of prey—the gyrfalcon and the snowy owl—are essentially creatures of the Arctic tundra. Many people consider the gyrfalcon the most magnificent of all falcons. It lives on the tundra, in the stunted woodlands, and in mountainous areas up to at least 3000 feet in mainland Arctic Europe, Asia, and North America, and on the islands of Greenland and Iceland. Although this falcon usually breeds north of the timberline, its range extends a short distance south into the great zone of coniferous forest called the *taiga*. The highest population densities are probably found near coasts. Totally flat ground does not suit it, but even a small crag suffices for a nest site, especially one that stands above a gorge or river valley. The area required to maintain a pair of gyrfalcons is large, varying from 25 square miles to 400 square miles. The birds may not use the whole of such an immense range during the summer, but may need to scour it all in winter to find enough food. The gyrfalcon flies very fast and usually low, probably catching most of its prey by pursuit rather than diving upon it. Ptarmigans form about 90 per cent of the gyrfalcon's food by weight, but it also feeds on seabirds and on ducks and geese. Its color phases range from very dark, almost black, to almost pure white. The white-plumaged birds predominate in the high Arctic, especially in northern Greenland.

Unlike the gyrfalcon, the snowy owl is always basically white, although the female is more heavily flecked and barred with dark brown than the male, which may be pure white. This owl lives on both low and high tundra, preferring

Common redpolls breed in the Arctic and subarctic of North America, Greenland, and Eurasia, particularly where spruce, dwarf birch, alder, and willow grow. In the high Arctic, these small birds choose ravines and rocky slopes. The paler Arctic, Hornemann's, or hoary redpoll breeds still farther to the north, even reaching northern Ellesmere Island, and stays there throughout the winter.

Two species of game birds occur in the Arctic: the willow ptarmigan (above left) and the rock ptarmigan (above right). Both species are mainly white in the winter, and largely brown or gray in the summer. This gives them some protection against birds of prey and predatory mammals, for the white color blends with the snow in winter, and the brown or gray merges with the tundra vegetation at other seasons of the year. Ptarmigan are protected against the cold by an extremely dense plumage that even covers their feet.

hummocky, rolling country. Its thick, dense plumage is an ideal insulator, extending over its feet to the tips of its toes, and almost hiding its hooked bill. The snowy owl is active mainly during daylight hours, perching on the ground or on such vantage points as posts or buildings, watching for its lemming prey before launching into a slow silent glide. Sometimes, however, its hunting takes the form of a swift dash after a victim, and snowy owls sometimes vary their diet with an Arctic hare or eider duck.

Periodically, populations of snowy owls increase dramatically. Usually this occurs about every four years, coinciding at least in part with a similar increase in the numbers of lemmings. Somehow, the sight of lemmings scurrying across the snow in spring determines how many eggs a female snowy owl will lay that year. In a year when few lemmings are visible she may lay only four eggs, but when they abound the number may climb to nine. The balance between the number of young snowy owls and the number of lemmings is fairly critical for the birds, because researchers have found that a pair of snowy owls and their offspring can eat up to 80 lemmings a day. When the population of these rodents crashes, the inflated owl population is faced with a serious food problem, and during the winter many of the birds may be forced to seek food far to the south of their normal range. This type of sudden migration is known as an *irruption*.

The Pleistocene glaciations must have pushed the ranges of all Arctic birds southward, and presumably they have recolonized the region only in the past few thousand years. Little is known about the species that inhabited the Arctic before the last ice age, but it is reasonable to suppose that freshwater aquatic birds such as the loons or divers were among the first to move north when the ice began retreating.

The common loon or great northern diver now breeds in northern North America, Greenland, and Iceland. American birds winter in Florida, Texas, and northern Mexico, and the others move south to spend the winter in western Europe. These large birds dive and swim magnificently, but can only shuffle about awkwardly on land because their legs are short and set far back on their bodies. They fly strongly with rapid wing beats, but are unable to take off from land; even from water they require a run of up to 1200 feet. Underwater, loons swim by paddling their webbed feet, descending under duress to depths of about 240 feet and remaining submerged for several minutes. Their normal feeding dives, however, are usually shallower and more brief. Loons in general come ashore only to nest. The site is usually no more than a yard or so from the water's edge and the nest varies in the different species from a scrape or depression in the ground to a crude heap of mud and weeds. After they have hatched, the chicks at first swim feebly, and prefer to travel on their parents' backs. At two weeks old they can dive, however, and at 10 or 11 weeks young divers can fly well.

The smallest of the family, the red-throated loon or red-throated diver, has the northernmost distribution, breeding from northern Ellesmere Island (west of north Greenland) to the southern limit of the tundra in North America, and in comparable regions in Eurasia. This bird needs a shorter takeoff than other loons, which means that the red-throated loons can hunt and breed on pools too small for their relatives. Moreover, this is the only diver known to be able to take flight from land.

Two species of grebes are found in the Arctic: the Slavonian or horned grebe, and Holboell's or the red-necked grebe. Both nest in the shallows and reed beds of quiet pools and rivers, and winter along the coast in sheltered bays and estuaries, migrating only far enough south to escape the worst of the northern winter. They are excellent swimmers, using their feet for propulsion. Unlike some diving birds, grebes steer with their feet instead of their tails, which are too short for this purpose. In further contrast

The gyrfalcon (above) is often considered to be the most magnificent of all the falcons. It flies very fast, usually quite close to the ground, and probably catches its prey—mostly ptarmigan—by chasing it rather than by diving on it from above.

The female snowy owl (right) is more heavily flecked and barred with dark brown than the pure white male. The snowy owl prefers hummocky, rolling tundra for breeding, and builds its thinly lined nest on the ground (below). It is active during daylight, and the number of young birds is quite closely related to the abundance of lemmings, which are its principal source of food.

The eerie, lonely, ringing cry of the loon is the epitome of the Arctic wilderness. The loons (or divers, as they are known in Europe) are large, solitary water birds, and the Arctic species include the red-throated loon (above) and common loon (left). They breed on tundra lakes and spend the winter at sea. Loons swim magnificently and are the most efficient of all the diving birds, on land, however, they are clumsy and awkward, for the legs are set far back on the body. They fly strongly with rapid wing beats, but the effort of taking off is enormous. Only the red-throated loon can take off from land; the common loon must do so from water, and even then it needs a run of up to 1200 feet. Underwater, the loons swim by means of their webbed feet, occasionally going as deep as 240 feet and remaining submerged for several minutes, covering perhaps as much as 500 yards before surfacing once again.

to many water birds, grebes' feet are not webbed; instead, each toe is separately fringed by a stiff, horny flap, or lobe. Grebes dive well, performing a neat little jump before slipping beneath the surface in search of insects, crustaceans, and mollusks. They also feed on water plants and seaweeds, but the two species mentioned eat fish only in winter, whereas fish form a major part of the diet of some other grebes and of the divers.

Swans, geese, and ducks are perhaps the most typical birds of the Arctic tundra in summer. Swans are the largest of all waterfowl, and three species nest in the Arctic. The whooper swan and Bewick's swan nest on the Eurasian tundra, and the whistling swan breeds on the North American tundra and *muskeg* (a northern type of tussocky bog) between Hudson Bay and Alaska. All three species migrate south for the winter. Largely because they are heavy—the whooper, at about 35 pounds, is near the upper weight limit for a flying bird—these creatures make hard work of getting airborne and need a long run over land or water. For landing they seem to prefer water, which presumably places less of a strain on their legs. Once in the air, swans though graceful, are not very maneuverable, and collisions with man-made objects such as overhead wires tend to be common and fatal. All swans are vegetarians. They feed on dry land by grazing on grass, and in water by upending to nibble aquatic vegetation. When feeding on

Two species of grebe live in the Arctic: the Holboell's or red-necked grebe (above left) and the horned or Slavonian grebe (above right). They nest in the reedy shallows of pools and rivers, and spend the winter in sheltered bays and estuaries along the sea coasts. Unlike those of most species of water birds, their feet are not webbed, but the separate toes are fringed by horny lobes.

The graceful swans are birds that most people associate with ornamental waters in north temperate lands. They are the largest of the waterfowl, and the tundra species include the whooper swan (above) and Bewick's swan (below). Because they are heavy birds—the whooper swan being near the limit for a flying bird—they need an unusually long run for taking off from land or water.

ponds, lakes, or rivers they remain in shallow water, because they dive rarely.

There are two natural groups of true geese: the "gray" geese of the genus *Anser* and the "black" geese of the genus *Branta*. All are confined to the Northern Hemisphere, and typically breed in Arctic and subarctic regions. The so-called "gray" geese include the pink-footed goose, white-fronted goose (of which there are four subspecies), lesser white-fronted goose, emperor goose, and three species of snow geese. The smallest snow goose is Ross's goose, which breeds in a single small area in north-central Canada and totals perhaps about 16,000 pairs. Like this species, the greater snow goose is pure white except for black primary wing feathers. It breeds on the coast of northwest Greenland, on Ellesmere Island, and on other nearby islands, but its main Old World breeding area is in north-eastern Siberia, on Wrangel Island off the Chukotski Peninsula. The island is now a reserve where between 120,000 and 130,000 pairs nest each year. Unlike the snow goose just mentioned, the lesser snow goose appears in two color phases: the white phase, in which the plumage closely resembles that of the greater snow goose, and the blue phase, in which the lower part of the neck, the breast, and the wings are usually a dark slate-gray. Both phases breed on the Arctic islands and mainland of northern Canada from Hudson Bay westward, and also in north-eastern Siberia, but the blue phase is commonest at the eastern end of this range and is spreading rapidly westward. It seems likely that the white phase will disappear altogether before many years have passed.

It might be expected that the nesting grounds of such large birds would be well known, but in fact the breeding area of Ross's goose was unknown to scientists until 1938, and no one yet knows where the Tule goose, a subspecies of white-front, hatches and raises its young.

Occupying flat coastal islands and high slopes of inland Arctic Europe, Asia, and North America are the various subspecies of the brant or brent goose. This is the smallest and darkest of the "black" geese, about the size of a mallard. Brants are more maritime than other geese, feeding along the coasts by day and by night. They are also strongly gregarious, and outside the breeding season flocks frequent stretches of coast where eelgrass grows. A close relative is the barnacle goose, so named because people once believed that the migrating geese appearing each autumn had hatched from barnacles. (For food purposes the geese were thus ranked as fish, which conveniently meant that Roman Catholics might eat them on fastdays when meat dishes were banned by the Church.) There are three populations of barnacle geese. One group breeds in eastern Greenland and winters in western Scotland and Ireland; another summers in Spitsbergen, wintering in Norway and Scotland's Solway Firth; and the third population breeds in the USSR, on Novaya Zemlaya and the western Siberian islands, and winters in Holland. Barnacle geese are very gregarious and largely nocturnal, moving over salt marshes, estuarine grass fields, tidal mudflats, and small grass-topped islands to feed. They breed colonially, building nests on ledges on steep cliffs, in river gorges, or sometimes out on the open tundra.

The most striking of the "black" geese is the red-breasted goose, a black, white, and chestnut bird somewhat smaller than a mallard. This species breeds on the Siberian coastal tundra and winters on the steppe surrounding the Aral Sea and the southern part of the Caspian Sea.

Because three quarters of all Arctic birds are adapted to aquatic or wetland habitats, it comes as no surprise to find that ducks are well represented. Indeed, only the perching ducks and stifftails are absent. The "dabbling ducks" of the Arctic include the mallard, green-winged teal, pintail, European widgeon, and baldpate or American widgeon. Dabbling ducks usually do not dive but take their food in shallow water by upending. They also feed on vegetation and small aquatic organisms floating on the surface, and eat grass and stubble on land.

The eiders, on the other hand, are essentially marine ducks. The common eider is a rather large bird that breeds all around the Arctic along coasts and occasionally inland by lakes. King eiders and spectacled eiders nest even farther north, and in the winter move south only far enough to keep clear of the sea ice. Steller's eider is restricted to the far north Pacific, and is seldom found south of Alaska. As with most other ducks, female eiders line their nests with down from their breasts. This eiderdown is a very soft, efficient insulator and it is harvested commercially. For this reason the birds are strictly protected in Iceland, the center of the industry. The dark ducks called scoters are also diving ducks, and maritime outside the breeding season.

They nest on beaches and small islands in the high Arctic, and winter off the coasts of the Atlantic and Pacific oceans.

Other Arctic ducks include the harlequin (the male of which has bizarre white markings) and the old squaw (or long-tailed duck), both of which breed on the tundra all around the north polar regions. There is also the red-breasted merganser, a hook-billed fish-eating bird that nests in the Arctic Circle in both hemispheres.

Many shore birds visit the Arctic; some nest nowhere else, but all of them depart for the winter. Here we shall briefly consider three of the most typical species: the ringed plover, the sanderling, and the red phalarope. The ringed plover is a stout little brown-and-white bird with a black breast band and a black mark across its eye. It is common and widely distributed in the Canadian Arctic in summer and also occurs in the breeding season in Britain, Greenland, Iceland, Scandinavia, northern Russia, and Siberia. In these places it haunts low coasts, and appears inland on gravelly plains and along the banks of rivers and lakes. It runs fast, pausing briefly to feed. The ringed plover crouches nervously at the approach of danger, and is skilled at feigning injury to distract attention from its young.

The tamest of the waders is the sanderling, a plump, extremely active little bird that resembles a clockwork toy as it runs along a beach chasing retreating waves. Sanderlings breed on the tundra in Canada, Greenland, Spitsbergen, and Siberia, and they are migratory visitors to the coasts of most continents, although rather scarce in the western Pacific. Mollusks and crustaceans form the bulk of their food, but in summer they also eat insects and some plant material.

The small, sandpiperlike phalaropes possess a number of peculiarities that separate them from other shore birds. In breeding plumage the female is more ornate than the male, and in the breeding ceremonies it is she who takes the initiative. On the other hand the male carries out most of the incubation of the eggs and brooding of the young. Phalaropes have dense ducklike plumage on their underparts, providing a raft of trapped air on which they float, and therefore they ride high in the water. Their toes are slightly webbed, and rows of flattened scales on the sides of each toe give them a fringed appearance. Phalaropes feed partly on land but they swim well and have a habit of picking food off the surface while they are spinning around in the water. Occasionally they upend like ducks, but they cannot submerge. The red phalarope, known as the gray phalarope in Britain, breeds all around the Arctic, almost exclusively north of the Arctic Circle. The females are the first to arrive at the breeding grounds, where they display in small groups, establish territories, and court the males. In the winter these birds migrate south as far as the Falkland Islands and New Zealand, living on the open ocean and feeding on such planktonic delicacies as small fish, crustaceans, and jellyfish.

Among the true seabirds of the Arctic those of the auk family are especially prominent. The common murre (or guillemot), Brünnich's murre (Brünnich's guillemot), and the black guillemot are broadly similar in appearance, being sharp-billed black-and-white birds that are predominantly black in summer, with extensive amounts of white plumage in winter. The common murre

The call of the wild geese as they head north to their breeding grounds in the spring is one of the most evocative sounds in nature. Arctic species include the lesser snow geese (left), white-fronted geese (above), and barnacle geese (below). The lesser snow goose has two color phases, both seen in the photograph: one is pure white except for the wingtips, the other a blue phase, with dark slate-gray neck, breast, and wings. The barnacle goose once used to be thought to hatch from barnacles.

spends most of its time in coastal and offshore waters, and breeds in dense colonies on cliff ledges and on the flat tops of isolated stacks in the North Pacific, the North Atlantic, and adjacent parts of the Arctic Ocean. Brünnich's murre feeds farther out to sea, but the smaller black guillemot (which is also known as the tystie or sea pigeon) is found much nearer the shore. This bird is often found among rocky or well-wooded islands, where it nests singly or in small scattered groups in holes or under boulders. It is almost circumpolar but is replaced in the Bering Sea and North Pacific by the pigeon guillemot. The razor-billed auk (or razorbill) is a sociable bird and is often seen perching on the same ledges as murres, which it rather resembles.

The Arctic's smallest wintering seabird is the dovekie or little auk, about the same size as a starling. This breeds in vast colonies in holes among rocks in Arctic Canada, Greenland, Ice-

land, Novaya Zemlya, Spitsbergen, and Franz Josef Land. It migrates over short distances, and during severe gales dovekies may be blown inland, or "wrecked." All auks have a very direct flight, and dovekies seem to whirr along on their short wings—short because they are adapted for swimming as well as for flying, and water is thicker than air. Dovekies seem to swim low in the water and to be sinking by the stern.

A puffin is usually quite unmistakable, with its short stumpy form, big head, and large parrotlike bill. The common puffin or sea parrot is the smallest of the three Arctic species, and inhabits the North Atlantic and nearby Arctic Ocean. It breeds around the Arctic coasts except apparently in Siberia and Alaska, where it is replaced by the horned puffin and the tufted puffin. Puffins are burrow-nesting birds, sometimes taking over the burrows of rabbits or shearwaters. However, they often dig their own holes (especially in the rabbit-free Arctic), using their bills as pickaxes and their webbed feet as shovels to throw the loosened earth backward. Although a puffin may pick up feathers, blades of grass, shells, and small bones, and take them down its burrow, the single egg is often laid on bare earth.

Skuas, gulls, and terns are quite closely related and are placed together in the family Laridae. Skuas (or jaegers) are far more oceanic than gulls and are essentially piratical predators and parasites, chasing and harrying other seabirds and forcing them to disgorge their last meal. This event often happens in flight, and the skua is usually fast enough to catch the food dropped by the luckless victim before the food hits the water.

The largest skua (called simply the skua, or the great skua, or bonxie) is one of the few birds with a bipolar distribution. One subspecies nests on the coastal tundra or moorlands of Greenland, Labrador, Iceland, and the Faeroes, Shet-

Male king eider (lower birds in this picture) seen in spring display plumage, in the high Arctic. The relatively few species that dwell near the poles face little competition, and—like the king eider—may be present in very large numbers. The twig nests of the eiders are thickly lined with warm down, and are hidden among low vegetation that shelters them from the keen northern wind.

lands, and Orkneys, and two others breed wherever suitable nest sites are available along the edge of the Antarctic continent and its islands, and on the coasts of southern South America, southwestern Africa, and the South Atlantic islands. Some scientists believe that the species originated in the north and then colonized the south. There, it evolved away from the more typical skuas and became more gull-like, subsequently moving back (perhaps fairly recently) to the north. However, there is some evidence that the northern form has been extending its range southward during the past half-century at the expense of its more northerly populations. Perhaps history is repeating itself.

Of the three remaining skua species, the long-tailed jaeger (or long-tailed skua) breeds the farthest north around the Arctic, although its population is relatively small and scattered. Curiously, for a seabird, its chief source of food is the lemming. The pomarine jaeger (or pomarine skua) and the parasitic jaeger (or Arctic skua), both breed along the coasts and on the islands of the polar basin. In all three species, dark, light, and intermediate color phases occur, rather like those of the gyrfalcon.

Gulls are long-winged birds with a graceful flight. Most kinds are white with gray or black

Above: the ringed plover. This small wader nests around the north polar region on Arctic coasts, and migrates southward in the winter. It is a nervous bird but is nevertheless adept at feigning injury to distract a predator's attention from its young.

The red phalarope (known as the gray phalarope in Europe) breeds right around the Arctic, north of the Arctic Circle. In the winter it migrates as far as the Falkland Islands in the South Atlantic, and to New Zealand—a long journey for a small bird.

The murres, or auks, are common seabirds of the Arctic. The common murre (above) nests in dense colonies on cliff ledges, and feeds in offshore and coastal waters. The dovekie or little auk (below) breeds in immense numbers in holes among rocks in Arctic Canada, Greenland, Iceland, and other places in the Arctic Circle. In winter it migrates as far south as Britain.

backs and wings. Five species are truly Arctic in character, but at least six others have been seen in the Arctic from time to time. The glaucous gull is the largest Arctic gull. It occurs in offshore waters, and along coasts and estuaries around the Arctic Ocean, breeding on all suitable cliffs, stacks, and islands. In the winter it migrates south to the British Isles, the Baltic, Japan, and California. Ross's gull, the smallest Arctic gull, ranges throughout the region, and often perches on ice floes and glaciers. The swampy tundra of northeastern Siberia is its breeding ground. Circumpolar in range, the ivory gull is the least known of the other Arctic gulls. It breeds only in the high Arctic, and its plumage is totally white. People usually see this species on the fringes of pack ice, and it nests on rugged, ice-bound coasts. Ivory gulls reportedly follow polar bears, scavenging on their dung and kills, but the birds are also known to catch small marine organisms.

Terns are more lightly built than gulls and two difficult-to-distinguish species of these slender "sea swallows" probably breed in the Arctic. The Arctic tern certainly nests there extensively, and the common tern—not in fact a common Arctic bird—breeds in Arctic Norway and may nest in the White Sea area and northern Canada. During the northern winter, the common tern migrates at least as far south as the Falkland Islands in the South Atlantic. The Arctic tern, however, is a more truly bipolar bird. It spends the northern summer breeding in the Northern Hemisphere up to at least 82° North, and as the Arctic winter approaches it flies south close to, or even beyond, the Antarctic Circle. The species was first seen in the Antarctic in the Weddell Sea in 1904, a sighting suggesting that at least some Arctic terns make an annual round trip of about 22,000 miles. But the longest proved migration was that of an individual tern banded (ringed) on 5 July 1955 at Kandalsch Sanctuary on the White Sea and recovered at Fremantle, Western Australia, on 16 May 1956. This tern must have made a one-way flight of at least 13,000 miles.

The Arctic tern probably sees more daylight than any other living creature. Birds arrive at the northernmost nesting sites when the midnight

The clownlike common puffin or sea parrot (right) carries fish crosswise in its bill to its young. On land, puffins walk upright with an apparently dignified air; they fly strongly with rapidly whirring wings, and are also excellent swimmers and divers.

sun is shining, and the sun does not set all through their nesting season. For at least two months of their stay in the Antarctic the situation is similar. Thus for about eight months of the year many Arctic terns experience 24 hours of daylight every day, and for the remaining four months have more daylight than darkness.

Arctic terns that visit the Antarctic add to that region's narrow variety of bird life. The southern counterpart of the Arctic tern is appropriately called the Antarctic tern. Antarctic terns breed on the Antarctic Peninsula and on many of the Antarctic and subantarctic islands, their nests being no more than slight scrapes in the ground, lined with blades of grass if these are available. During the breeding season the adults are noisy and defend their territories vigorously, swooping and diving upon intruders while screaming and chattering with apparent rage. Like all of Antarctica's birds, the terns survive by exploiting the sea. They do not normally submerge, but scoop up fish as they swoop low over the water. However, those breeding at South Georgia have been seen to plunge into offshore beds of kelp in search of their prey.

The only gull usually seen in the Antarctic is the Dominican or kelp gull. This species is the southern counterpart of the lesser black-backed gull of the Northern Hemisphere. It is circumpolar and has one of the most extensive ranges among seabirds, extending from Marguerite Bay west of the Antarctic Peninsula, to Lobos de Tierra Island off northern Peru. Birds of this species remain permanently in places where climatic and other conditions permit, for example in South Georgia and Macquarie Island; otherwise they fly to more hospitable regions for the winter. The Dominican gull feeds on the eggs and young of other birds, and on carrion, fish, mollusks, sea urchins, and crabs. In the Falkland

Jaegers (skuas) are known to be scavengers and to attack smaller birds to make them disgorge fish, but are unlikely to have killed the royal penguin they are eating here. The brown skua is one of the few species that breeds in both polar regions.

By breeding in the Arctic and then wintering in the Antarctic, the Arctic tern (above) probably sees more daylight than any other bird. Its migration is the longest known of any bird species.

The largest Arctic gull is the glaucous-winged gull (below), found over offshore waters and along coasts around the Arctic Ocean. Its winter homes are as far south as Japan and California.

The Dominican gull is the southern counterpart of the lesser blackbacked gull of the Northern Hemisphere. It has one of the most extensive ranges known among seabirds.

Islands these gulls have been seen breaking open mollusk shells by dropping them on tarmac roads, which become littered with broken shells.

The only birds without webbed feet known to have reached the Antarctic continent unaided are two species of shore birds—the sheathbills mentioned earlier. Sheathbills take their name from the horny, saddle-shaped sheath covering the base of the upper bill. Describing these curious creatures, possibly anatomically linked with the waders and gulls, one observer wrote that they have "the general appearance, gait, and flight of a pigeon, the beak and voice of a crow, the habits of a wader, yet dread the water, and the pugnacity and familiarity with man of a raptor." The sheathbill has an especially strong flight and is occasionally found 300 or 400 miles from the nearest land or even the nearest ice floes. Sheathbills are in general quarrelsome, sociable, and bold, unafraid even to walk on the backs of elephant seals. They are primarily scavengers, and will feed on such materials as seals' afterbirths, eggs (which they steal even from beneath incubating cormorants), small fish, mollusks, algae, and lichens.

By contrast with the clumsy-looking sheathbills, the magnificent long-winged wandering albatross is the world's most splendidly graceful bird—at least when seen in soaring flight. It covers mile after mile of ocean with no effort beyond a slight tilt of its wings. This is the largest seabird and, indeed, the largest of all flying birds, if we go by wingspan rather than by weight. The condors and some swans are heavier but their wings are shorter. The wanderer's wings measure at least 11 feet from tip to tip, 13 feet in some cases. Weight probably averages between 17 and 20 pounds, of which the skeleton accounts for as little as two and a half pounds. The wandering albatross roams the westerly wind belt that circles Antarctica, ranging from the Tropic of Capricorn to 60° South, and occasionally flying over the pack ice, especially in the northern part of the Weddell Sea.

The earliest arrivals at the breeding grounds on the subantarctic islands are generally males. Territories are established, and courtship displays take place when the females arrive. Pairing birds of both sexes puff out their chests, stand on their toes, stretch their wings, raise and spread their tails, and touch their bills together. As if this display were not impressive enough, there is much clattering of bills, and shrieking and groaning. The single egg is laid in a nest that looks rather like a miniature volcano; it is a steep-sided construction of grass and twigs, about two feet high and two feet wide at the top. Both parents take turns in incubating the egg for a total of about 80 days, and because skuas find albatross eggs and young chicks desirable additions to their diet, the parents guard their nest very carefully. After about five weeks of brooding the chick is as heavy as the adults, although still covered by thick gray down. At this stage the adult birds abandon it, and for three months it lives on the nest, using up its reserves in a state of near hibernation. Finally the chick ambles off the nest and begins to try its legs and wings. However, a long period of ground-training follows before it is ready for flight. Occasionally, indeed, chicks are still on the nests when the adults return for the new breeding season. These chicks are unceremoniously pushed off.

The young birds seem to spend several years at sea before attempting to breed. Eventually they return to their original home and for several seasons go through the courtship rituals and form pairs, but produce no eggs. A wandering albatross may be eight or nine years old before it rears its first chick. Because they mature slowly, lay only one egg a year, and yet manage to survive as a species, the wandering albatrosses must live very safe lives above the stormy deep.

Albatrosses fly best in moderately high winds, and find difficulty in calms. In fog, they are said to settle on the water. From land or water a long takeoff run is almost always necessary except when the birds can launch themselves from a hillside, although once they are airborne they soar at up to 60 mph and can easily cover 300 miles a day. Alighting on dry land presents another problem for these big seabirds. They descend breast-first and often somersault before coming to rest.

Although the wanderer is the most spectacular albatross, the black-browed albatross or mollymawk is the commonest in the southern oceans. Albatrosses seldom pass from the Southern to the Northern Hemisphere because they cannot normally cross the doldrums (the windless regions astride the equator). A few birds do somehow pass this barrier, however, and since 1963 more than 20 albatrosses have been recorded in British and Irish waters alone. One individual was a black-browed albatross that lived on Bass Rock in Scotland's Firth of Forth in 1967 and 1968, disappearing to sea in the winter.

Albatrosses form one of the four families of petrels, an old order of seabirds that may be distantly related to the penguins. The fulmars, prions, gadfly petrels, and shearwaters belong to

The only bird without webbed feet to reach the Antarctic continent is the sheathbill (above). It is a quarrelsome, pugnacious scavenger, and is disliked by most visitors to the region.

Aggressive during the breeding season, the Antarctic tern (above) will attack intruders near its nest, screaming and chattering. It swoops low over the water to scoop up its food.

a second petrel family, and great numbers of these birds breed in the Antarctic. The giant petrel, the largest member of the family, is often seen flying with the albatrosses. Whalers and sealers called it the "Nelly," for some now-obscure reason. Its alternative name of "stinker" is much more obviously fitting, because an intruder at the breeding site is likely to receive an accurately directed jet of oily liquid squirted from the nostrils of a disturbed bird. This substance, secreted by the petrel's stomach, has a sickly, fishy, cloying smell that clings to splashed clothes for a very long time. At sea the giant petrel feeds on krill, fish, and squid, but it also eats on land, taking offal from whaling operations, and chicks from penguin colonies.

The smallest members of this second petrel family are the prions (or whalebirds), which nest in burrows that they excavate in the cold, peaty soil. Because they are heavily preyed upon by skuas, the prions dig chiefly at night, although most attacks occur when they are entering or leaving the burrows and they can usually outfly their predators. Prions feed in two ways. One method involves sitting lightly on the water with their wings held up, paddling with their feet, and pecking at food, or briefly diving. But a prion may also catch squid, small crustaceans, and marine worms by "hydroplaning"—running along the surface with wings outstretched and head submerged so that its open bill plows through the plankton-rich water, trapping plankton and straining water out through the bill's comblike sides. (In fact, this feeding method is basically that used by the much larger flamingos and by the mighty whalebone whales.)

To many people, the snow petrel is one of the most beautiful birds in the world. It has a more southerly range than any other bird except the emperor penguin, and is invariably associated with ice.

The remaining two petrel families include the diving petrels and the storm petrels. Diving petrels are the Southern Hemisphere's ecological equivalent of the king auklets of the North Pacific. Both are small and short-winged, and have a whirring flight. Diving petrels live in the stormy seas of those southern latitudes aptly named the Roaring Forties and Furious Fifties: regions of high winds, lowering clouds, and heavy, driving rain. These petrels feed below the surface of the water, seemingly flying through the swells, swimming with their wings, and steering with their feet.

Wilson's petrel (or Mother Carey's chicken), best known of the storm petrels, is a little black bird often seen near the wake of ships as it flits over the waves with rapidly beating wings. It usually feeds from the surface, apparently skipping along on tiptoe with wings fluttering, but it can also dive to a depth several times its own length. Wilson's petrel ranges over the southern

The seas south of the Antarctic Convergence are rich feeding grounds for many seabirds. The large birds above are wandering albatrosses, whose wings span about 11 feet. Also present with them are three giant petrels and two pintado petrels.

oceans, and during the Antarctic winter as far north as California, Labrador, the British Isles, Arabia, India, New Guinea, and Japan.

Another striking petrel is the snow petrel, one of the most beautiful birds in the world. The adults are pure white—so white, in fact, that they make snow and ice appear dirty by comparison. Their immaculate appearance is enhanced by the black bill, eyes, and feet. Strangely, their underdown is black too, and it seems likely that this hidden, dark, inner layer may help to warm the birds by trapping the solar radiation that penetrates their feathers. Snow petrels have a more southerly range than any bird except the emperor penguin and the Antarctic skua, and they are invariably associated with ice. Individuals have been found breeding on nunataks nearly 200 miles inside Antarctica.

Finally, we move on to the penguins, the birds that for most of us typify wildlife in Antarctica. In fact, penguins breed not only there but along the coasts of Southern Africa, South America, and Australia. One species even breeds astride the equator on the Galápagos Islands off Ecuador. Penguins, then, are widely distributed around the Southern Hemisphere, and by no means confined to ice and snow. Possibly they evolved on the temperate southern shores of the prehistoric southern supercontinent Gondwanaland. Fossils about 100 million years old back up this notion, and also show that the ancestors of modern penguins could fly. The penguins' colonization of polar waters seems to have been relatively recent, but emperor and Adélie penguins, in particular, have become fully adapted to their difficult environment. The chinstrap, the macaroni, and the southern race of the gentoo penguin live and breed on the Antarctic islands but do not nest on the continent, whereas the king, the rockhopper, and the northern race of the gentoo penguin breed on the subantarctic islands. The plumage of all these Antarctic penguins is very dense, with up to 70 feathers per square inch on the back. Their heads and bills are relatively small, and the Adélie's bill is well feathered. Because penguins feed on marine life, diving ability is very important. But the larger the bird, the more energy it expends in diving and the more food it needs in order to replace that energy. On the other hand, a big penguin can store more oxygen, and therefore dive deeper or for a longer time than a small penguin. Thus, the emperor penguin can apparently spend up to three minutes submerged, whereas the smaller species manage no more than one minute or so.

By far the largest of present-day penguins is the emperor, which stands about four feet high and weighs up to 90 pounds. This is probably the only bird that almost never sets foot on dry land, and no other vertebrate breeds so far south. Adults gather on newly formed sea ice in the Antarctic autumn (April or May), and, without going through the formality of claiming territory or building a nest, each female lays a single egg, weighing a little less than one pound, in late May or in June. The male begins incubation, carrying the egg on his feet and under a flap of abdominal skin. Incubation altogether takes from 50 to 60 days, and during most of this time the male fasts, losing about one third of his weight in the process. Meanwhile the female travels out to sea to feed. After some weeks she returns, completes incubation, and rears the chick. The male, in his turn, goes to sea, and the

The fussy, inquisitive, charming Adélie penguin is a great traveler and is capable of finding its way home over extremely long distances. It is often apparently reluctant to enter the water, perhaps because it is sometimes eaten by the predatory leopard seals.

Emperor penguins, seen here crossing the tide crack near Shackleton's hut at Cape Royds, Ross Island, are by far the largest of present-day penguins. They are probably the only birds that almost never set foot on dry land, and breed farther south than any other vertebrate. In order to be able to rear their young at the best possible time, emperor penguins begin their nesting on the bare ice in midwinter. The male incubates the single egg on his feet for about two months, losing about one third of his weight during this time.

female soon follows. In October or November the emperor penguin chicks move to the ice edge, and when the ice breaks up into floes the young birds are borne north. This is a critical time, for the floes must persist long enough for the young to complete their molt and assume seaworthy plumage. The main enemies of the adult emperor penguin, as for all Antarctic penguins, are the leopard seal and the killer whale, but blizzards take a heavy toll of the young.

Lacking the apparent dignity of the emperor and king penguins is the comical and charming Adélie penguin. This inquisitive and seemingly cheerful and fearless little bird is the favorite of all who have visited the Antarctic. As soon as a ship ties up to fast ice, one or two Adélies appear and solemnly regard all that goes on. Soon, a small group is inspecting the ship and the invading humans from close range, trotting about or slowly tobogganing from one vantage point to another. Although they are quite easily handled, Adélies are difficult to catch because of the speed with which they can move over snow and ice, and also because of the way in which they use bills, flippers, and clawed feet in self-defense.

The Adélies' range includes the coasts of the Antarctic continent and adjacent islands, but they wander north to South Georgia, Heard Island, and Macquarie Island. Indeed, there is proof that they are great land travelers too. In January 1958 tracks seemingly made by Adélies were found about 400 miles from the United States' Byrd Station and heading east toward the South Pole. At this point they were at least 184 miles from the Antarctic coast. The observers followed the tracks for a mile or so and discovered that the birds had alternately walked and tobogganed. It appeared to be a purposeful journey and not just a random wander.

Adélies are capable of homing over vast distances, as an experiment begun in December 1959 showed. Five males were taken by air from Wilkes Land to McMurdo Sound—about 2400 miles. By the following October, three of these birds were back in Wilkes Land, having traveled at the remarkable average rate of some eight miles a day. Other Adélies, repeatedly moved 35 miles, always returned to their starting point.

These penguins return from the sea to their land-sited breeding rookeries in October, when they may have to walk many miles across sea ice to reach them. There are probably more than 100 rookeries, ranging in size from just a few birds to several millions. On arrival the Adélies build their nests of small stones, which they steal from their neighbors as necessary. It seems that these pebbles prevent the eggs from rolling into the meltwater pools. Despite the intense cold of their surroundings the penguins contrive to keep their eggs at the high temperature of 92.7°F. (Biologists established this fact by sawing an egg in half, inserting a telemeter thermometer, and returning the doctored egg to the parent bird for incubation.)

Penguins and some other Antarctic seabirds can live in vast numbers, thanks to the wealth of food that they find in the sea. Our next chapter examines the polar oceans themselves, and the network of life forms they support.

Adélie penguins return to the breeding rookeries (above) in October, when sea ice is still present, and they may have to walk or "toboggan" (left) many miles to reach them. On arrival they build their nests of small stones, which they may find for themselves or steal from neighbors. The function of the pebbles may be to prevent the eggs rolling into the very muddy surroundings. The Adélie penguin completes its breeding in three months, compared with the six or seven months required by the emperor penguin, and during this time the scene is chaotic. The air filled by the braying "Aark-aark" of hundreds, if not thousands, of birds, and the smell of rotting guano is overwhelming. But soon the birds depart to sea, the snow falls, and all is clean and peaceful.

The Polar Seas

Enormous areas of both polar regions carry ice and snow upon which men can walk, if they are hardy (or foolhardy) enough to do so. Indeed, since the middle 1950s expeditions moving over snow and ice have crossed the surface of Antarctica and journeyed over the top of the world from Alaska to Spitsbergen. Yet this last journey took place on sea ice and the explorers who crossed Antarctica first had to sail across an immense tract of chilly sea to get there. Oceans, then, are major features of both polar regions, although they differ in that one surrounds an island continent, whereas most of the other is hemmed in by land. Both polar oceans support only cold-water life forms. Yet the special character of each mass of water and its distance from the other help to explain why Arctic Ocean plants and animals are by no means identical

Sunset and moonlight over the pack ice off northern Greenland. As the long polar night begins, new ice will form between floes and bind them together until the Arctic Ocean surface is an ice mosaic, glowing in the light of the moon and stars.

with those found in the Southern Ocean. In this chapter we shall compare the coldest oceans in the world and their inhabitants.

Compared with the immense and very deep Atlantic and Pacific oceans, the Arctic Ocean is small. It covers a mere five and a half million square miles to the relatively shallow average depth of 4000 feet, although the sea floor plunges to an impressive 17,850 feet at the ocean's deepest point. Another special feature is, of course, the ocean's coldness. The temperature of surface waters exposed to the Arctic atmosphere can fall as low as 28.8°F before the waters freeze, because the salts dissolved in seawater reduce its freezing point below the 32°F at which fresh water turns to ice. Upon further cooling, however, ice forms on the ocean surface. At first this sea ice comprises small, very thin plates, but these may grow bigger until they form large floes up to 10 feet, or even 15 feet, thick. Much of the Arctic Ocean is covered throughout the year by pack ice (masses of such ice), which drifts in a roughly clockwise direction around the North Pole.

We saw earlier that the Antarctic Ocean, or Southern Ocean, as geographers usually call it, is formed by the southern parts of the Atlantic, Pacific, and Indian oceans. But the Southern Ocean's area depends upon the way in which it is defined. Most oceanographers place its northern limit at the Antarctic Convergence—the warm-water/cold-water boundary lying between 50° and 60° South. Here, surface waters flowing northward from the Antarctic continent dip beneath warmer, and hence less dense, subantarctic water flowing generally southward. The result is a sharp change in temperature and salinity

(the warmer and saltier water lying north of the convergence). The character of the marine life also changes. But not all oceanographers consider that the Southern Ocean ends at this boundary. Some regard the Southern Ocean as extending northward to about 40° South. This brings it as far north as the southern coasts of Africa and Australia, and approximately to the northern limit of the drift of Antarctic icebergs. In these latitudes yet another abrupt change of temperature and salinity occurs, this time along the so-called Subtropical Convergence. That boundary would give the Southern Ocean an area of about 29 million square miles, making it about six times as large as the Antarctic continent and more than one fifth the size of all the oceans combined. But for the purposes of this book, the biological barrier of the Antarctic Convergence is a more meaningful boundary.

During the southern winter the Southern Ocean freezes to a depth of between three and nine feet in a more or less unbroken band extending from 60 to 120 miles from the Antarctic mainland. Meanwhile, the coalesced floes comprising pack ice extend much farther (up to an average of 500 miles), and the pack ice covers altogether seven and a half million square miles, an area nearly half as great again as that of the entire Arctic Ocean. Of course considerable thawing takes place in summer, and the Antarctic pack ice eventually recedes to a relatively modest one million square miles. Even during the depths of winter, however, the sea ice is not an unbroken sheet isolating the ocean from the air above. Fissures form between the floes, sometimes extending for miles as wind or currents open up temporary *leads* (navigable lanes) and *polynyas* (areas of open water). Such open waters are vital to seals, whales, and penguins and other seabirds, allowing the mammals to surface and breathe, and permitting the birds to feed.

Polar marine life produces relatively simple food chains and exceedingly complex food webs, just like life on land. In the sea, as on the land, herbivorous animals feed on plants, and carnivorous animals feed on herbivores. Ultimately all marine animal life depends on the growth and abundance of plants. Especially important are minute single-celled plants making up the so-called *phytoplankton* that floats freely in the sunlit upper levels of the sea, providing food for the small, often larval, animals that form the *zooplankton.* This zooplankton includes the shrimplike crustaceans known as *krill*, a major food of whalebone whales, seals, and many birds, including the penguins. The leopard seal eats penguins, and both fall prey to killer whales. Obviously these relationships are greatly simplified—many other animals, including fish and mollusks, enter into the web.

The sizes of animal populations largely depend upon the amounts of food available, and thus it is important to discover what controls plant growth in the sea. Broadly speaking, the ingredients needed for successfully producing marine plants are adequate amounts of dissolved oxygen and carbon dioxide, heat, light, and nutrient compounds including nitrates and phosphates. Dissolved carbon dioxide (required for photosynthesis) and oxygen (required for respiration) are generally abundant because cold polar water

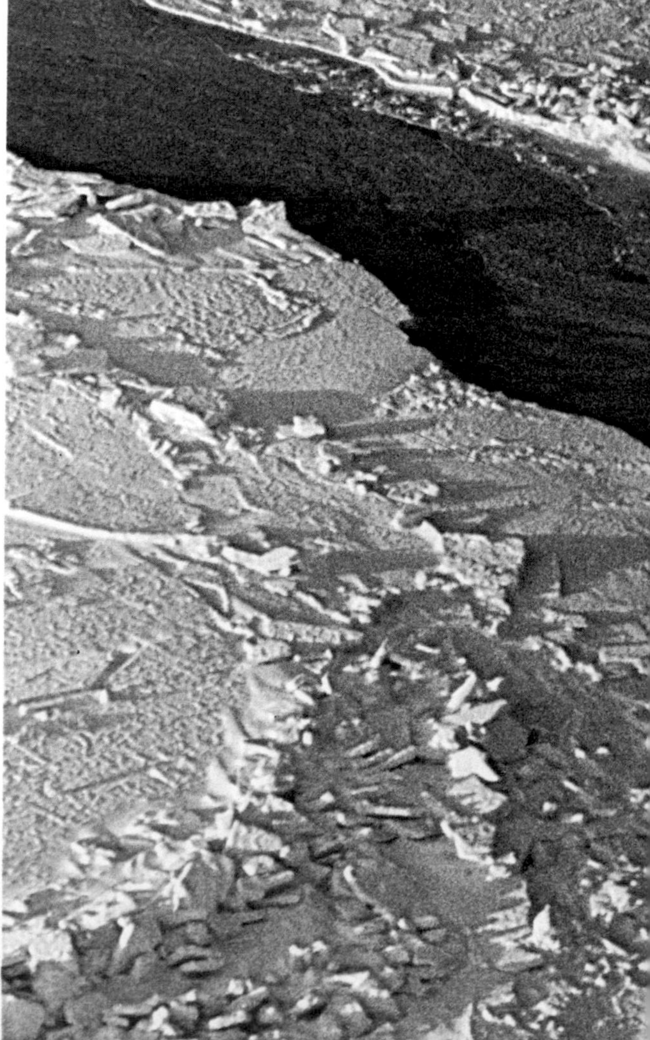

Aerial view of a lead—a strip of open water in the pack ice. Leads provide valuable breathing holes for sea mammals, especially whales. Belugas and other whales swim up and down the watery lanes—a routine that helps to keep the leads ice-free.

can hold more dissolved gases than warm tropical water. On an annual basis the relative lack of heat and light does not seem seriously to hamper plant growth, although low temperatures may slow the rate of growth, and low light intensities reduce the amount of biochemical activity in green plants during the long polar night. Nevertheless, in Arctic waters, the total amount of plant production is low. This is because of a shortage of nutrient compounds, not in the ocean as a whole but in its upper layers where conditions are otherwise suitable for photosynthesis. Nutrient salts released by the bacterial breakdown of dead plants and animals sink slowly to the ocean floor, and are effectively lost unless the water deep down is somehow raised and mixed with surface water. In Arctic waters such mixing takes place only to a very limited extent. Scientists believe that during the winter the density of Arctic Ocean seawater is fairly uniform from top to bottom; there is little relatively heavy water to sink and little relatively light water to rise. In spring and summer, less saline, and hence lighter, water from melting ice floes and glacial runoff floats on top of the more dense oceanic water, again forming a stable system in which little mixing occurs.

Conditions in the Antarctic are quite different. Considerable mixing takes place here because waters flow away from the continent along the surface and are replaced by waters upwelling from the nutrient-rich depths. As they rise, these replacement currents carry nutrients toward the surface. Consequently in Antarctic waters marine plant production can continue at a high level for much of the year. In the Arctic, similar conditions do occur in places, but only at the seaward ends of glaciers in Greenland and Spitsbergen, and these upwelling currents are too small to affect the general circulation.

Thus, in the Arctic, plant production is limited at all seasons, although in early spring there is a phytoplankton "bloom," when the usually clear water becomes a green soup. Although Antarctic marine plant production is considerable almost all the year, here, too, there is a dramatic spring

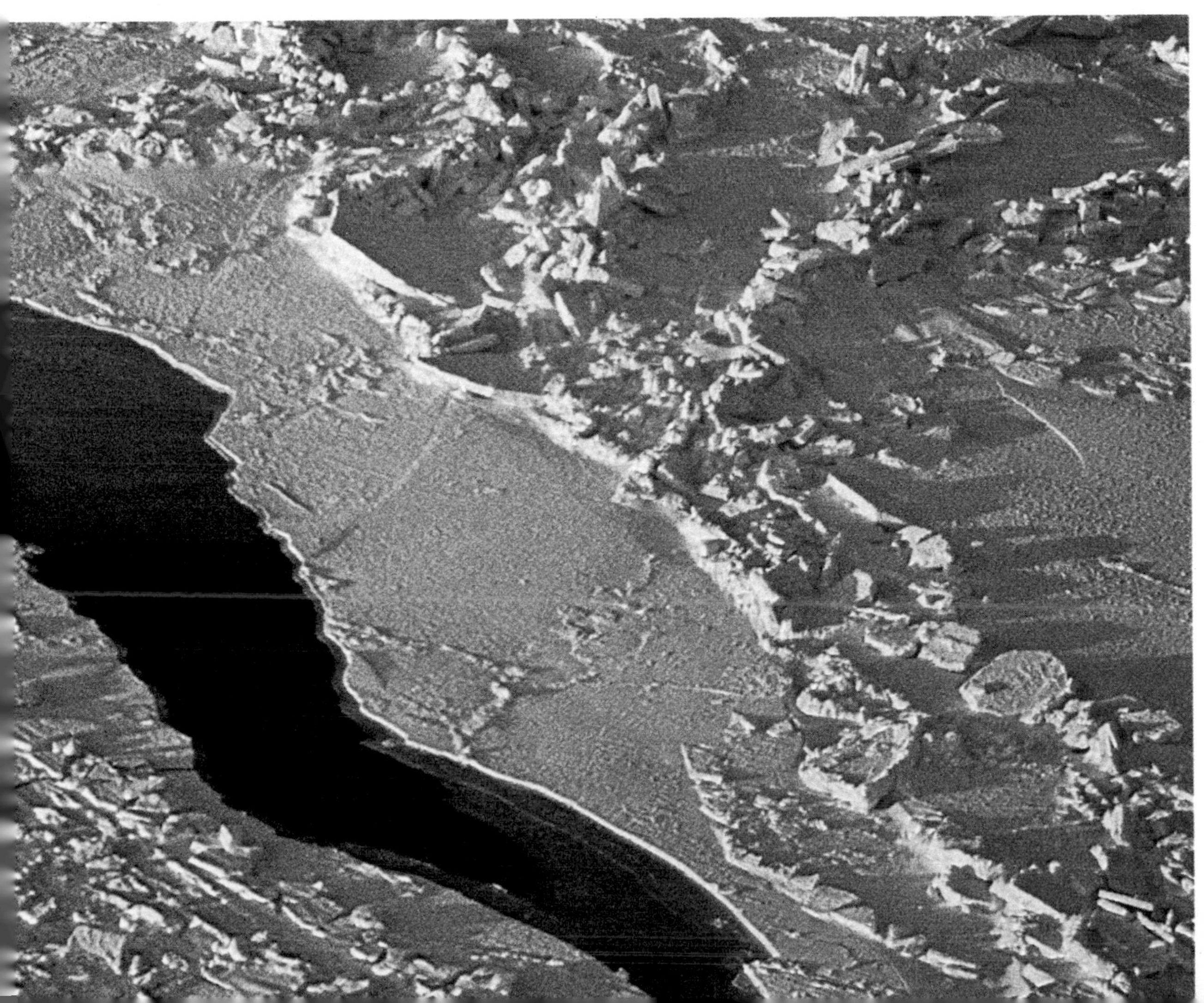

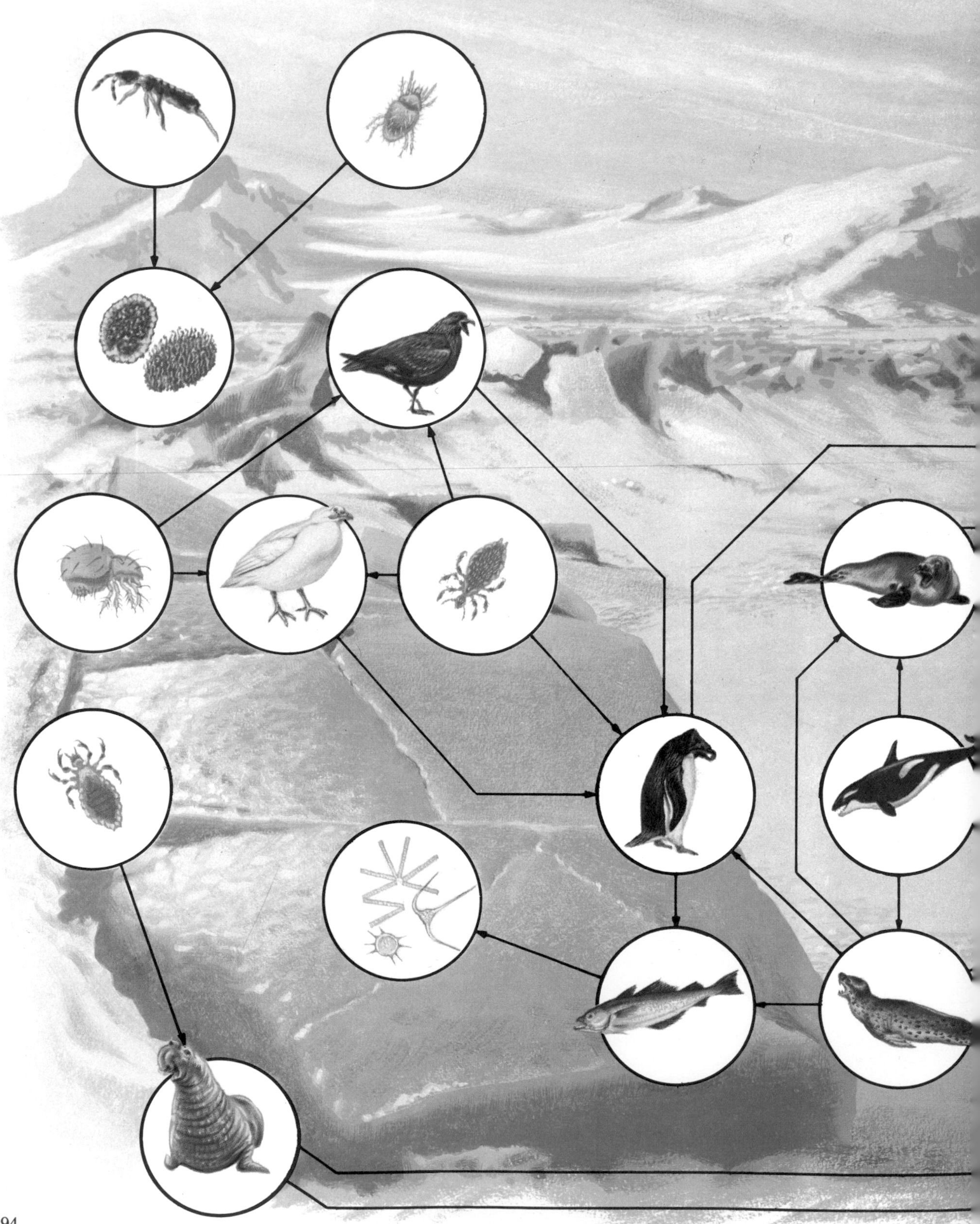

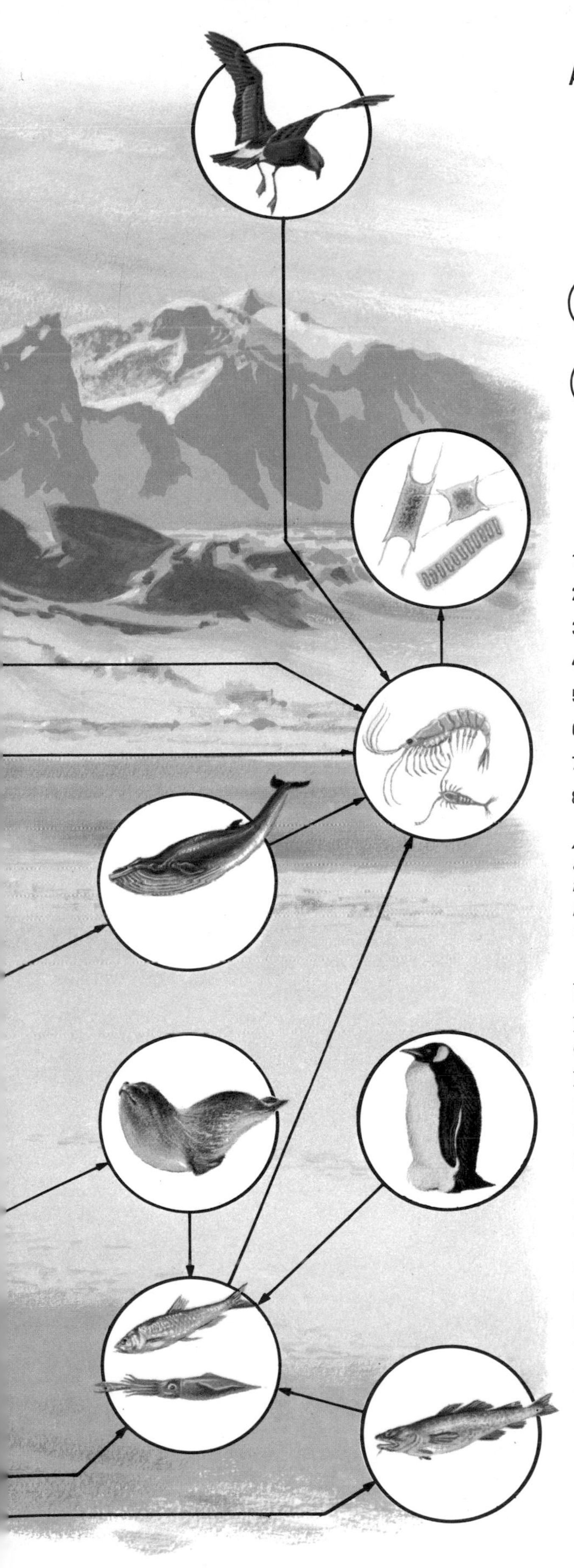

An Antarctic Food Web

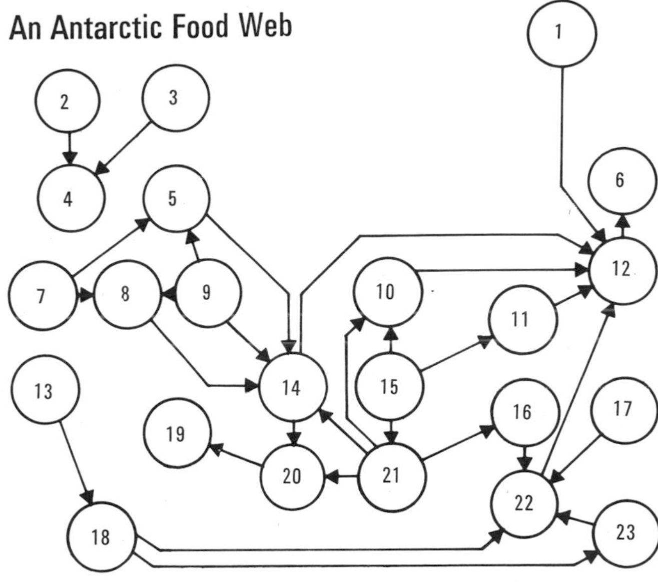

1 Petrel
2 Springtails
3 Mites
4 Moss and lichen
5 Skua
6 Phytoplankton
7 Parasitic mites
8 Sheathbill
9 Biting lice
10 Crabeater seal
11 Rorqual
12 Krill
13 Sucking lice
14 Adelie penguin
15 Killer whale
16 Ross seal
17 Emperor penguin
18 Southern elephant seal
19 Diatoms
20 Fish
21 Leopard seal
22 Small fish and squid
23 Large fish

Apart from springtails and mites, which are supported by mosses and lichens, almost all Antarctic animals ultimately derive their food from the sea. Even the mites that parasitize sheathbills form part of a food chain that passes via the Adélie penguins to the krill and the phytoplankton that abound in the Southern Ocean.

bloom. The herbivorous zooplankton of both polar regions also shows an annual population explosion, known as a "flush." This flush takes place a little later than the bloom of the phytoplankton upon which the herbivorous zooplankton grazes, and the flush of the carnivorous zooplankton follows soon after.

Most planktonic organisms float in the upper levels of the open ocean, but in polar regions there is a rich growth on the underside of ice floes. This often shows up as a yellow or brown stain caused by masses of minute silica-shelled diatoms. These are single-celled plants that, in the Antarctic at least, make up 99 per cent of the phytoplankton. Somewhat surprisingly, enough light penetrates the ice to allow photosynthesis during late spring and summer, although layers of snow accumulating on the floes in fall and winter drastically reduce the light entering

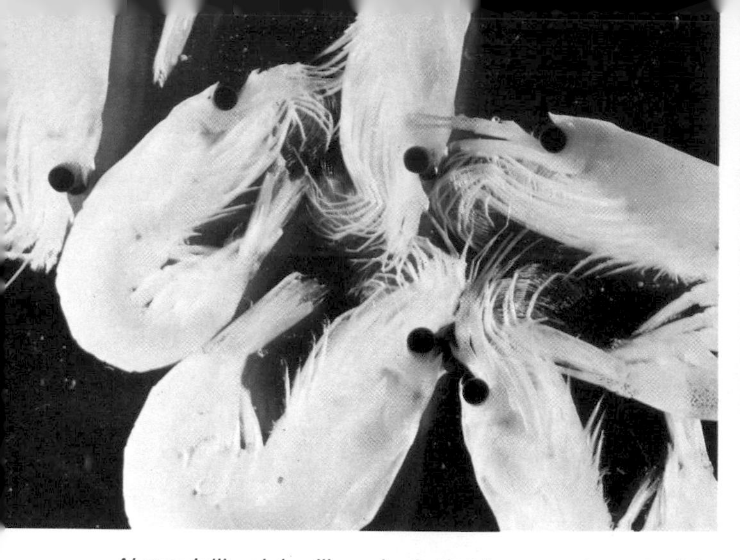

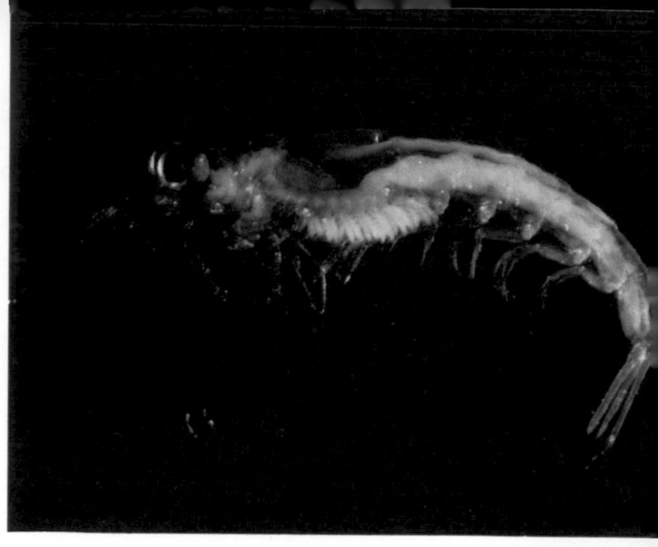

Above: krill—shrimplike animals that form a major part of the diet of a variety of polar fish, sea mammals, and seabirds. Euphausia superba *(left) abounds in Antarctic waters.* Meganyctiphanes norvegica *(right) is its Arctic counterpart.*

Below: a mass of creatures trawled from the sea off South Georgia. Such samples show that ice fish, crocodile fish, octopuses, sea urchins, sea whips, sea spiders, sea fans, sea anemones, and hydroids live on or very near the subantarctic seabed.

the sea, and thus curb phytoplankton growth.

In terms of sheer mass of animal material produced, krill is probably the most abundant type of invertebrate in polar seas. The term *krill* (from a Norwegian word for baby fish) was coined by whalers to describe the pink shrimplike crustaceans that form the main food of the whalebone whales. There are two quite distinct kinds of krill. In the North Atlantic, one-and-a-half-inch-long *Meganyctiphanes norvegica* forms the diet of the whalebone whales and the herring. In Antarctic waters, however, krill is composed of a two-inch-long species of *Euphausia*. One of the commonest species in this genus is *Euphausia superba*, which forms the principal food of the rorquals, especially the blue and fin whales. But squid, crab-eater seals, and many seabirds including penguins also eat it. This krill species cannot survive in water warmer than 39°F, and thrives only south of the Antarctic Convergence.

Another type of crustacean common in polar seas and particularly characteristic of the Arctic Ocean (but also found in oceans everywhere) is the small copepod *Calanus*. This creature is colorless, transparent, and about the size of a grain of rice. Normally it swims vertically but, like all the members of the plankton, it is at the mercy of the currents. It also tends to sink, but reduces the rate of sinking with the aid of two long, feathery antennules, one projecting from each side of its body.

Phytoplankton and zooplankton both occur in heavy concentrations in polar seas where upwellings, and possibly winds, bring nutrients to the surface. In these dense patches, whales, seals, fish, and seabirds congregate to feed.

So far, we have considered marine invertebrates that live on the surface. Polar regions also include animals that live in the mid-depths of the sea. Squid are a notable example. These somewhat octopuslike mollusks usually range from 10 inches to six feet long, but giant species up to 46 feet are known, and there is some evidence that even larger forms exist. A squid seizes and eats fish and other prey with the help of 10 thick tentacles equipped with suckers and hooks, and a mouth resembling a parrot's bill. But squid themselves fall prey to larger predators. These mollusks' horny bills are often found in the stomachs of toothed whales, especially the sperm whale, and whales have been caught whose heads and bodies bore scars as big as dinner plates, caused by the suckers of immense squid with which the whales had grappled.

On polar coasts the seabed is usually almost bare of the multitude of life forms found in more temperate tidal zones. This largely reflects the scouring effect of ice. Much of the Antarctic margin is locked in the frigid armor of permanent fast ice—so called because the ice is fastened to the shore. Elsewhere, ice seasonally covers beaches. Thus polar tidal zones are often frozen solid. Then, too, icebergs and floes that are blown ashore grind along the seabed, destroying any developing plant or animal life. Farther out to sea, however, below the level down to which ice forms, the polar seabed supports a rich population of invertebrates, including sponges, sea lilies, sea cucumbers, sea squirts, starfishes, and sea urchins. The seabed

mollusks are represented by snails, limpets, mussels, and clams, which form an important food of carnivorous animals including the walrus.

As with other living organisms, there are proportionally fewer species of fish in polar waters than in warmer seas, although their total numbers are moderately rich. Most kinds of fish live at or near the bottom and only a few species inhabit the surface layers. But some Arctic species are *cryopelagic*: that is, they swim actively in mid-water but at some stage in their lives they make their homes close to drifting or fast ice. Many such fish are members of the cod family. These include the aptly named *polar cod*, which has been caught close to the North Pole. It is a small, dark-silvery fish that often takes up residence in fissures, cracks, and cavities in the underside of ice floes, especially when these are thawing and breaking up in the spring. In this habitat it feeds on the zooplankton and takes refuge from its enemies, for the polar cod in turn is preyed upon by other fish and by the harp, ringed, and bearded seals, the white whale, the narwhal, and many seabirds. Polar cod are nonetheless abundant, migrating and collecting in enormous premating schools in fall and winter. Another common Arctic fish is the glacial cod, found off the coast of Greenland and in deep water close to the North Pole. It feeds on plankton, and forms part of the diet of seals.

Several species of fish breed in subarctic rivers, but spend a good part of their lives in the Arctic

Left: rows of Arctic char hung out to dry at Disko Bay, west Greenland. Arctic char provide a valuable winter food for Eskimos, who catch them as they reenter rivers from the sea, grown large and plump on plankton and young capelin, a herringlike fish. The adult Arctic char breed in lakes, and their young thus start life in fresh water.

Below: Antarctic cod, a fish found south of the Antarctic Convergence. Antarctic cod have big heads and relatively small, flattened bodies. They can survive in water below the temperature at which blood usually freezes, by concentrating products in the blood. These lower the freezing point of the blood, and thus act as antifreeze.

Ocean. One of these is the Arctic char, which has the most northerly distribution of any freshwater fish, ranging throughout the northern polar region. The Arctic char occurs in lakes and rivers of the northern North American and Eurasian mainlands and on Greenland, Iceland, and many other Arctic islands. Mature specimens spawn in fresh water in the fall, and the young fish move downstream to the sea when they are about six to eight inches long. By this time they may be five to seven years old, for their growth is slow. Arctic char are carnivorous and feed on a wide variety of animals, such as insect larvae, small mollusks and crustaceans, and fish, including the polar cod and their own species.

Only about 100 species of fish are known from south of the Antarctic Convergence (out of a possible world total of some 20,000 species) and a high proportion belong to the order of perchlike fishes, the Perciformes. Many are cryopelagic and among the most interesting are the nototheniids, or so-called "Antarctic cod." These fish have large rounded heads that become proportionally even larger with age, and relatively small bodies that are somewhat flattened from side to side. One of the largest species is *Notothenia rossi*, which is found off South Georgia and the South Shetland Islands and measures up to one yard or more long. These fish are opportunistic feeders, taking almost anything edible, even—in waters close to scientific bases—potato peelings and cabbage leaves. Like the Arctic char, they grow rather slowly, and first breed when they are nine or 10 years old and have reached a length of about 14 inches. In May, each adult female lays 12,000 to 14,000 eggs, averaging one eighth of an inch in diameter. The fertilized eggs sink to the bottom, but the young fry swim to the surface after hatching and lead a planktonic existence until they are about three and a half inches long. They then descend to the sea floor, where they spend much of their adult lives.

Fish pose one of the most intriguing puzzles of polar survival because, particularly in winter, the waters they swim in are close to the seawater freezing point of 28.8°F and below the freezing point of blood. Why don't fish freeze solid? It seems that there is more than one answer, depending on the fish. In some species, for example in *Notothenia neglecta* and *Notothenia rossi*, there is an increase in the concentration of sodium, potassium, and chloride ions, and of urea, in the circulating blood in winter. This increased concentration lowers the freezing point of the blood in much the same way as putting salt on a wet pavement in winter lowers the freezing point of water and prevents it turning to ice. In other species, such as *Trematomus borchgrevinki*, there is an increase in the winter concentration of blood glycoproteins (protein-containing carbohydrates). These substances seem to confer increased resistance to freezing in supercooled conditions—conditions where the seawater is below the temperature at which the fish's body fluids freeze. But this is not the whole story, for some kinds of fish swimming happily around in supercooled conditions soon become drowsy and die if they are touched with a small ice crystal. Indeed, they freeze solid. Some fish, however, spend a large part of their life in contact with ice, and scientists believe that ice crystals in water are lethal to such fish only if the ice comes into direct contact with their extremely sensitive gills. Certainly, such fish seem well designed to prevent this happening. Their respiration is so slow that they can get enough dissolved oxygen by taking water into the mouth through almost closed lips. By doing this they probably prevent ice crystals from being carried by the current of water flowing over their gills. Perhaps it is a matter of "he who keeps his mouth shut, lives longest."

Antarctic waters are also the home of the strange "bloodless" fish. In fact, this name is incorrect, for such fish do possess blood, but it lacks red blood cells. (These cells contain hemoglobin, the pigment that transports oxygen in the blood of all other vertebrates.) Some of these "bloodless" fish, or ice fish, as they are also called, grow more than a yard long. They have large heads and slim bodies, and appear superficially normal except that their gills are creamy-white instead of pink. Scientists have found that the oxygen-carrying capacity of the blood of these fish is only about one ninth that of the blood of other Antarctic fish, and is about what one would expect of the plasma (the fluid part) of normal red blood. The Antarctic seas are well aerated, and it may be that ice fish not only take up dissolved oxygen through their gills, which are well developed, but also absorb some oxygen directly through their scaleless skins. Although some species of "bloodless" fish appear to behave sluggishly, others are known to feed on krill, a very active crustacean. This means that the fish must be active swimmers, but just how they obtain enough oxygen to produce the energy required remains a mystery.

Above: an ice fish, also called a "bloodless" fish because, unlike all other vertebrates, it has no oxygen-absorbing red blood cells. Ice fish seem to take up all the oxygen they need through their gills and possibly directly through their skin.

Two glimpses of life upon the Antarctic sea floor. Left: sea anemones and other invertebrates. Above: sea spider, a long-legged, spiderlike invertebrate of a type common in polar seas. The seabed off icebound polar coasts lacks such life forms.

Seals and Whales

Early polar explorers were astonished by the vast numbers of seals and whales that swarmed in the seas of the Arctic and Antarctic regions. Centuries of hunting mean that these marine mammals are far less numerous now, but thousands still make their homes at the bitterly cold ends of the earth, where they are remarkably successfully adapted to life in icy conditions.

It is not known just how marine mammals arose. The oldest known fossils of these seals and whales are of animals already adapted for life in the water and give no clues to link them with land-living ancestors. But biologists believe that the whales probably arose as an early specialized offshoot of the placental mammals, the group to which most living mammals belong. The oldest known whales, the extinct Archaeoceti ("Ancient Whales"), have been found in Middle Eocene deposits about 50 million years old; the toothed whales are known from late Eocene rocks about 40 million years old, and the whalebone whales appear in Middle Oligocene rocks dating from 35 million years ago. The earliest known fossils of seals are much later in origin, and date from about 16 million years ago, in the Miocene epoch.

Seals are *the* amphibious marine mammals of the polar regions. They can haul themselves out on land or on ice floes to rest and must leave the water to molt and to breed, and of course—unlike fishes—have to surface to breathe. Nevertheless, they are equally at home in the water; it is there that they spend most of their lives and obtain their food. They all have a streamlined, torpedo-shaped body to assist their movement through water, and all are expert swimmers and divers. Like its landbound ancestors, a seal possesses four limbs, each with five digits, but its limbs have evolved into flippers.

Seals form three distinct, but quite closely related, groups. These are the true, or "earless," seals (Phocidae); the eared seals (Otariidae), which include the fur seals and the sea lions; and, in a family by themselves, the walruses (Odobenidae).

All seals show some characteristics of carnivores, and it is probable that they arose from terrestrial carnivore ancestors. Many scientists believe that the true seals are descended from otterlike ancestors, whereas the eared seals arose from something more bearlike or doglike, but some authorities think that both originated from the same stock and then diverged. The eared seals owe their name to the fact that they possess external ear flaps *(pinnae)*, although in a greatly reduced form; the true seals have none at all. Also, the palms and soles of an eared seal's flippers are hairless, whereas those of a true seal are well furred. The major structural difference between these two groups, however, involves the limbs, especially the hindlimbs, and affects their mode of locomotion both in and out of the water. The true seals swim by means of a strong sculling action of the hindlimbs, which trail and cannot be rotated forward even when the animals are out of water. Coupled with the fact that these seals have relatively short forelimbs, this means that true seals cannot raise their bodies clear of land, ice, or snow, and on such surfaces hump themselves along using their forelimbs only, dragging their hindquarters behind them. In water, a true seal's forelimbs are generally used only for steering and balancing.

By contrast, in the eared seals the forelimbs appear to provide most of the motive power for swimming, and the hindlimbs remain relatively inactive in the water. But on land these seals can rotate their hindlimbs forward under the body, lifting it clear of the surface. In this way they can "walk" on all four limbs and even manage a clumsy-looking gallop.

The three subspecies of walrus are put into a family by themselves because they possess characteristics of both the other groups. Like the eared seals, walruses can rotate their hindlimbs forward; but, like the true seals, they lack external ear flaps. Most of their swimming effort comes from the hind flippers, but they swim less powerfully than some other seals.

Despite the fact that seals are found in all the oceans, most live in areas of cold water. With a few exceptions, notably the three species of monk seal, even those of temperate or tropical seas live in areas affected by cold currents. But the vast majority of both species and individuals inhabit the food-rich seas in or near the icy, inhospitable polar regions.

The Pribilof fur seal lives in the northern Pacific Ocean. Amid much roaring and fighting, mature bulls establish territories at the same breeding grounds each year and, as the females come ashore to give birth, the successful bulls collect a harem of about 50 cows.

The bearded seal, named for its great profusion of whiskers, is an Arctic species that prefers shallow waters near coasts free of fast ice. It is not commercially important, but is of great value to the Eskimos for both its meat and its skin.

In general, the true seals inhabit the coldest waters, and their headquarters are in the Northern Hemisphere. Of the five species that live south of the equator, four (the Weddell, leopard, Ross, and crab-eater seals) breed on Antarctic ice. The remaining southern species, the southern elephant seal, is primarily subantarctic in distribution but sometimes breeds on fast ice at the southern edge of its range. Of the 13 species of true seals living north of the equator, six breed on Arctic ice: these are the ringed, bearded, hooded, ribbon, harbor, and harp seals.

The harp seal, named for the harp-shaped marking on its pelt, is a migratory species that inhabits the edge of the Arctic pack ice and the subarctic waters of the North Atlantic from Hudson Bay in Canada to Cape Chelyuskin in north-central Siberia. There are three separate populations, each with distinct breeding grounds (the White Sea, the Greenland Sea north of Jan Mayen Island, and the Newfoundland area) and there seems to be no movement between them.

The harp seal's environment is a cold one, and any warm-blooded animal living in such circumstances must have a very effective means of regulating its body temperature and of insulating itself against a massive heat loss. Seals have a greater problem in this regard than most other mammals because they spend a great deal of time in water, and although the air temperature in both polar regions may be considerably lower than that of the surrounding seas, the cooling effect of water far exceeds that of air.

Seals solve this problem by a combination of biological devices. The first of these involves their resting metabolic rate (the rate at which bodily processes, such as respiration, digestion and so on, occur when an animal is at rest). A seal's resting metabolic rate is higher than that of most land mammals, which means that it can make up for heat losses more readily. But if this were the only adaptation enabling seals to keep warm, one would expect their metabolic rate to increase as their surroundings cool down. In fact, within certain limits, the metabolic rate of all seals so far studied remains stable during such cooling, and the harp seal is metabolically un-

The harp seal is a migratory species found in the Arctic Atlantic Ocean. The adult has a large, harp-shaped dark-brown band straddling the back and shoulders. The pups are born with a woolly white coat that very soon molts to a short gray one.

affected even when submerged in ice water. Thus other temperature controls must also apply.

Chief among these is an efficient system of insulation, well illustrated in the harp seal. Like all true seals, the harp seal is covered with short hair that, on land at least, provides some insulation by trapping a layer of air next to the skin. In water, however, the hair soon becomes wet and loses what little insulating property it has when dry. But all adult harp and other true seals have a layer of fat (or blubber), approximately three inches thick, beneath the skin except over the head and appendages. Blubber is an effective insulator on land and in water, serves as a food reserve, and gives buoyancy. In addition, these seals can if necessary shut off the outer blood vessels from the main circulation, allowing only enough blood through to feed the tissues and prevent them from freezing. Thus, when a seal is in cold water, or when part of its body is in contact with ice, it does not attempt the impossible task of keeping all of its body equally warm, but merely closes some of the outer blood vessels. This results in a massive temperature gradient through the blubber, so that while the seal's interior remains at, say a cosy 98.6°F, the surface of the skin may be as low as 35.6°F. This lower outer temperature drastically cuts heat loss to the surroundings and allows the seal to conserve body heat without increasing its metabolic rate. In warm surroundings the blubber might lead to overheating, but should the seal need to lose excess heat, it merely dilates the outer blood vessels and thus lets heat escape from the surface of the skin. In water, flippers are generally kept at a lower temperature than most of the rest of the seal's body, so that energy is not wasted in keeping them warm. Nevertheless, the flippers are freely supplied with blood vessels and are thus also useful as radiators.

Harp seals spend most of the year migrating and feeding in loose herds on the edge of the Arctic pack ice, following it northward in summer and southward in winter. The adults' diet consists primarily of capelin, herrings, and haddock, and the pups feed on planktonic animals and small fish.

Seals display various patterns of social organization but, apart from a few of the true seals, most kinds are strongly gregarious at least during the breeding season. For example, toward the end of winter, harp seals gather in pupping rookeries that vary from five to 100 square miles in extent and may contain nearly 6000 seals per square mile. The bulls indulge in courtship displays, cavorting in leads between the ice floes, swimming madly in circles, "porpoising," and shooting straight out onto the floes to attract the females' attention. Although two males may fight over the attention of a single female, these seals are monogamous and there is no harem formation. (This is so for all true seals except the gray and elephant seals, whose harems average 10 and 20 females respectively.) Mating occurs about two weeks after the pups are born and takes place either in the water or on the floes. The harp seal's apparent gestation period is about $11\frac{1}{2}$ months, but implantation of the fertilized ovum in the uterus is delayed for some 10 weeks—a physiological device ensuring that pupping occurs at the right time of year. This is an important feature in seals, for, if their young are to have the best chance of survival, they must be born after the worst winter storms are over

The ringed seal is one of the commonest of the Arctic seals, and it occurs as far north as the pole itself. It is hunted by many Eskimos, who use the meat for food and the skins for warm winter clothing such as that worn by the child pictured below. This small seal is the favorite prey of the polar bear.

and be able to take to the water in spring or summer, the time when the krill on which they feed is most abundant. Among seals in general, pupping usually occurs in late winter or early spring (October through November in the Southern Hemisphere, February through April in the Northern Hemisphere), and the youngsters have a very short suckling period during which their mothers' rich milk permits them to grow at an incredible rate. As a further adaptation to polar conditions, most seals living in the coldest areas are weaned more rapidly than those in warmer climates.

Harp-seal pups are born on the floes from late February through early March. They are then about two and a half feet long, and weigh about 12 pounds. Like the newborn young of most species of seals, they have little blubber, but make up for this lack with a dense, woolly coat of long white hair that molts two to four weeks later, leaving a sleek, short-haired, gray coat more suitable for life in the water. Harp-seal pups are among the most precocious of infants, being abandoned by their mothers after only about two weeks. However, nursing during the first week is almost continuous, the youngsters wailing plaintively if left even briefly; by the time their mothers cease to give milk, the pups have increased their weight eightfold. Much of this rich milk goes toward the development of blubber. The pups then starve during molting, and when they take to the water at about one month old their weight has halved to some 50 pounds. Male harp seals reach sexual maturity in eight years, females in six.

The ringed seal, so called for the pale rings on its brown back, is one of the commonest Arctic seals. In some areas it is still the mainstay of the Eskimos' economy, providing them with skins for clothing, meat for food, and oil for lamps. It lives from the North Pole south to the subarctic, wherever there are open-water leads. This seal does not migrate and spends much of the winter under the ice, keeping the vital breathing holes open by constantly breaking any newly formed ice with its head.

The pups are born from mid-March through mid-April in ice-dens which may be natural or hollowed out by the mother. The newborn young are about 25 inches long, weigh 10 pounds, and are covered with a long, creamy-white fur that molts after two to three weeks. Suckling continues for almost two months. Mating appears

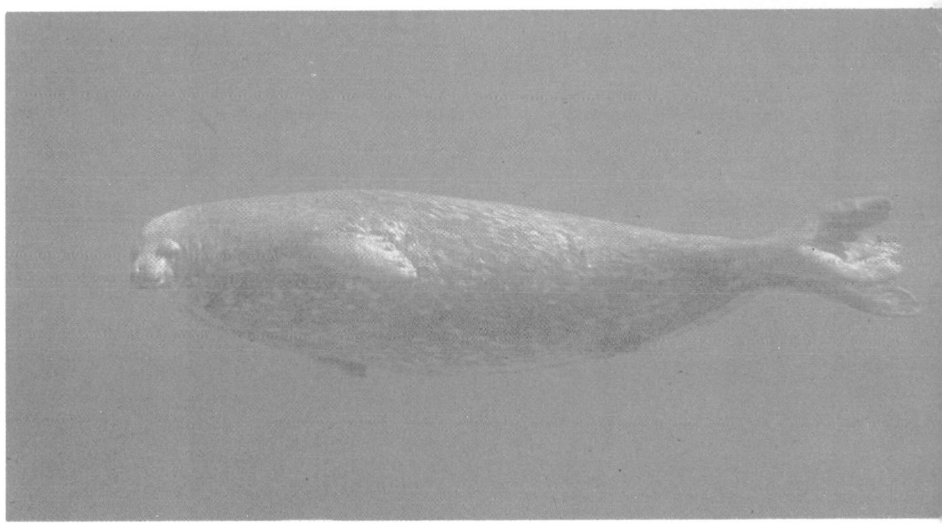

The Weddell seal is a large chubby creature that lives farther south than any other mammal except for man and his huskies. It does not migrate, and spends the whole winter under the fast ice (right), surfacing to breathe through holes it opens up with its incisor and canine teeth (above). How these seals find their holes again in the gloom under the ice is not known for certain, but their large brown eyes are well supplied with the photoreceptors responsible for vision in dim light. During the long winter night Weddell seals can be heard "singing" under the ice, and it is possible that they use the strange high-frequency sounds to locate their breathing holes.

to take place shortly after the females have given birth, but implantation is delayed for about 14 weeks. Both sexes reach sexual maturity after about seven years.

The ringed seal eats at least 72 different species of small fish and planktonic crustaceans, but during the molting season in June and July adults lie on the ice and do not feed. This seal's main enemies include man, polar bears, killer whales, and Arctic foxes.

Of the four other seal species breeding on Arctic ice, the bearded seal (brown with long whiskers) occurs from Greenland to Norway; the hooded seal (notable for its inflatable snout) roams among drift ice between Greenland and Spitsbergen; the harbor, or common, seal (brown with dark spots) occurs near Arctic, North Atlantic, and North Pacific shores; and the ribbon seal (brown with yellow bands) lives in the Bering Sea.

Apart from the southern elephant seal, comparatively little can be said about the true seals of the Antarctic, because little is known about them. In general, they are larger than their northern counterparts, but seem to live shorter lives. The Weddell seal, named for the English sealer James Weddell, is one of the most numerous, and apart from man and his huskies, no other

The gregarious crab-eater seal (above) is probably the most abundant of the various Antarctic seals, and is frequently to be seen basking on the drifting pack ice. By contrast, the large-eyed Ross seal (below) is a solitary animal that is seldom seen by man. It lives almost exclusively in heavy pack ice.

mammal lives so far south. It is most frequently seen on the coasts of West Antarctica and in the Ross Sea, where during the summer it lazes on the ice in small, scattered groups. Food consists chiefly of fish, but also includes sea cucumbers, squid, and crustaceans.

The Weddell seal does not migrate but spends most of the winter under the fast ice, breathing through holes that it makes by swinging its head from side to side and scraping away at the ice with its canine and incisor teeth. The resultant wear on these teeth is an important cause of death in this species, for seals with badly worn teeth cannot keep their breathing holes open and eventually drown. Just how foraging Weddell seals find their holes in the gloom under the ice remains something of a mystery, but their eyes are large and have many of the special photo-receptors that help them to see in dim light. Echolocation (finding objects by detecting echoes

bouncing off them) is also a possibility, because these seals "sing" underwater, producing a variety of high-frequency sounds; but their use of echolocation has yet to be proved.

Weddell-seal pups are born in September and October, and are covered in grayish, woolly fur. Their first molt occurs about two weeks later and lasts about a month, leaving the pups with an adult-type coat. Suckling lasts from six to seven weeks, and throughout this period the mother goes without food.

The crab-eater seal is the most abundant of the Antarctic seals, some estimates placing its population as high as 20 million. It is frequently found on the drifting pack ice and basking on the floes in the Weddell and Ross seas and along the west coast of the Antarctic Peninsula.

Adults of both sexes are slim creatures and can move rapidly over ice. Males are about eight and a half feet long and weigh 500 pounds,

The leopard seal (above) has a reputation for ferocity, which is not helped by its rather reptilian appearance and by its occasional habit of taking a penguin for dinner (below). It probably does not attack man unless provoked, however, and its main food appears to consist of various fish, squid, and krill.

The southern elephant seal is the largest of all seals, although the female (left, with her black, woolly-coated pups) is considerably smaller than the male. Its name is derived not only from its large size but also partly from the trunklike elongated nose of the male. This proboscis can be inflated during the breeding season and may help to amplify the bull's roar. These seals breed on many subantarctic islands and on parts of the Antarctic mainland, and breeding territories are established by the bulls before the cows haul themselves out to give birth. Tremendous clashes take place as the bulls fight for their territories, and they inflict seemingly terrible wounds on each other (below). Their weapons are their large canine teeth. As the cows come ashore, harems are formed, the largest and strongest of the bulls gathering around them the greater number of females.

females are slightly larger. Little is known of their breeding habits, because the cows give birth from September through November, when there is usually no one around to observe them. But it is known that the newborn pups are about four and a half feet long and covered with a soft, woolly, grayish-brown coat. Suckling probably lasts about five weeks, but could end after two weeks, when the first molt occurs.

The crab-eater's chief enemy is the killer whale, and many adult seals bear the scars of encounters with these predators. These seals do not eat crabs, as their name implies, but feed almost entirely on krill. A crab-eater swims into a mass of krill with its mouth open, then shuts its mouth and filters off the water through its three- to five-cusped cheek teeth.

Even less is known of the habits of the Ross seal, named for the Scottish explorer James Clark Ross, because this is a solitary animal seldom seen by man. The few records available indicate that the adult female is about eight and a half feet long, and the male a little larger (about the same size as the Weddell seal). The Ross seal is a chubby creature with noticeably prominent eyes, and a short, wide head that can be retracted into rolls of fat around the neck. Adults are known to molt in January, and during this period they fast. At other times their food consists of squid, fish, and krill.

The solitary leopard seal is a beast of ill repute, and has a somewhat reptilian appearance, with a slim body bearing leopardlike spots, a snakelike head, and many sharply pointed teeth. Its reputation as a ruthless killer is largely undeserved; it probably does not attack man unless provoked, and although it does eat penguins—skinning them by biting off the tail and feet and beating

The Kerguelen, or southern, fur seal (above) belongs, in fact, to the sea-lion family. It has a similar life history to the Pribilof fur seal of the north. Gross overhunting during the 19th century led to the southern fur seal becoming so scarce that its hunting was eventually abandoned as unprofitable. They are now protected on many of the Antarctic and subantarctic islands, and their numbers are increasing rapidly.

the body against the surface of the water—its main food appears to be fish, squid, and krill. Leopard seals are migratory and range widely to most of the subantarctic islands in winter. They are usually seen on the outer edges of the pack ice and some adults stay among the pack ice throughout the year. Little is known of their breeding habits.

The largest of all seals, the southern elephant seal, breeds on most of the subantarctic islands as well as on the mainland itself. Adult males are 18 to 20 feet long and may weigh nearly four tons. Full-grown females are considerably smaller: roughly about half the length and one quarter the weight of the males. This seal received its name not only because of its large size, but also because the adult male develops an elongated nose rather like a short "trunk," which in mature bulls is long enough to overhang the mouth. During the breeding season this proboscis can be inflated and probably helps to amplify the bull's roar, which can be heard for miles.

Bulls come ashore early in September to establish their breeding territories. This involves much bellowing and fierce fighting, a challenger rearing up and then lunging forward to slash at an opponent's throat and chest with his large canines, often inflicting terrible, though seldom fatal, wounds. Toward the end of the month the cows come ashore, and harems, usually consisting of 30 to 40 cows, are formed. The black, woolly-coated pups, born about a week later (from matings the previous year), are some four feet long, and weigh 80 pounds. Suckling continues for about three weeks, during which time the mothers fast and may lose as much as 700 pounds. Meanwhile, the pups are gaining weight at the incredible rate of about 20 pounds a day. The first molt starts at about 10 days of age and finishes about 24 days later. During the first month following weaning, the survivors use up their blubber reserves and then eat small crustaceans before taking to the adult diet of squid and fish. Despite their mothers' care, as many as half of all the pups die: some get trapped in the holes they have dug while wallowing in peaty soil; others are accidentally crushed by adults.

As a group, the eared seals favor cold, but not polar, waters. Their headquarters are in the Southern Hemisphere, but only one species, the Kerguelen fur seal, actually breeds south of the Antarctic Convergence. Only three of the 13 species of eared seals live north of the equator and none of these breeds as far north as the Arctic Circle, although both the Pribilof fur seal and the Steller's sea lion enter Arctic waters during the summer.

Best-known of the three eared seals occurring in polar seas are the Pribilof and Kerguelen fur seals (the third is the Steller's sea lion), both sought since the days of the early sealers.

Pribilof, or northern, fur seals breed in the North Pacific Ocean on the Pribilof Islands, on some of the Commander Islands, and near Sakhalin. During the winter, they migrate singly or in small groups southward to Honshu and California. The Kerguelen, or southern, fur seal is divided into two distinct subspecies, only one of which breeds south of the Antarctic Convergence, on islands such as Bouvet, Heard, and South Georgia, on the South Sandwich, South Orkney, and South Shetland island groups, and on the Kerguelen Islands, which actually straddle the Antarctic Convergence. This seal's distribution outside the breeding season is uncertain, but biologists believe it moves north with the shoals of krill upon which it feeds.

Both these species have the fur seal's characteristic thick coat, which grows in two distinct layers: an upper layer of long, coarse guard hair and a lower one of short, dense, velvety fur that serves as an insulator by trapping bubbles of air next to the skin even when the animal is underwater. Fur seals are so well insulated that on land in warm weather they suffer from overheating, and can then be seen panting and waving their flippers in the air like fans. They lack the true seal's main heat-control devices—the blubber is thinner, and they do not have such fine control over the blood supply to the skin.

The life histories of these two species are very similar. Both are polygamous, the bulls generally arriving at the breeding grounds first and establishing their territories with much roaring and fighting. As the females come ashore to give birth, harems are formed, each consisting of as many as 50 females in the Pribilof species. Those of the Kerguelen fur seal are smaller, generally averaging around five females. The pups are born shortly after the cows come ashore (November through mid-December in the south, June

The ungainly-looking walrus probably uses its tusks, which grow continuously throughout its life, for many purposes, including defense and raking over the seabed for shellfish. Both sexes have tusks, but those of the female are smaller.

through July in the north), and mating occurs a few days later. There is a period of delayed implantation of about 14 to 16 weeks in the Pribilof fur seal. Biologists believe that active pregnancy probably lasts about nine months in the Kerguelen species, but the duration of the delayed implantation is still uncertain.

At birth, Pribilof fur-seal pups are about two feet long, weigh about 12 pounds, and are clothed in coarse black hair. Their mothers suckle them for some three months. Newborn pups of the Kerguelen fur seal are about one and a half feet long and have a black, woolly coat that molts to a gray coat when suckling ceases in April. Almost the first thing a pup hears after birth is a characteristic high-pitched howl from its mother, which it answers with a bleat. This duet between mother and pup continues for some minutes, and thereafter a cow can distinguish between her offspring and all others by sound, although final identification is always made by smell. Females reach maturity after about two years, males after about three years, although few become successful harem bulls until they are at least six years old.

There are three races of walrus: the Pacific subspecies, found mainly in the Bering Sea and along the northeast coast of Siberia; the Atlantic subspecies, occurring along the Arctic coast of Canada from Hudson Bay to Baffin and Ellesmere islands, parts of the Greenland coast, Iceland, and the northwest coast of Siberia; and the Laptev walrus, which lives in the Laptev Sea and regions of the Kara and East Siberian seas.

The three races have similar habits and appear almost identical. Adult males are about 12 feet long and weigh nearly one and a half tons, whereas adult females rarely exceed 10 feet and weigh less than one ton. The scanty coat of reddish-brown hair becomes sparser as the animal ages, and these seals must rely almost en-

The whale's tail, with its two large, horizontally flattened flukes—seen here as a humpback whale dives beneath the surface of the water—is its main propulsive unit. Steering and balance are provided by the foreflippers. The tremendous size and power of these enormous marine mammals allowed them to roam the world's oceans virtually unmolested until man destroyed their supremacy.

The rorqual known as the minke whale, shown here, is exhibiting behavior characteristic of many species of whales. It is not certain why they jump out of the water in this way, but it is thought that they possibly do so in an effort to rid themselves of clinging barnacles, or perhaps use the resulting splash as a sound signal to other whales. On the other hand, they may do it just for their own pleasure.

tirely on their two-and-three-quarter-inch-thick layer of blubber for warmth. Both sexes have tusks, which are upper canine teeth that grow continuously throughout the walrus's life. A male's tusks may reach a length of one yard and a weight of 12 pounds, but a female's tusks are shorter and lighter. It is thought that the walrus uses its tusks in a variety of ways: as levers to help the animal haul itself onto an ice floe, as picks to keep breathing holes open, as stabbing swords for defense and for disemboweling the seals on which it occasionally feeds, and as rakes to scour the seabed for the shellfish that form its main diet.

Pups, born in April and early May, are about four feet long and weigh 100 pounds. At birth the coat is slaty-gray, becoming reddish-brown at the postnatal molt. Suckling continues for more than a year in this species and the weaned young remain with their mother for a further year or so.

We have shown that different seals are adapted in various ways for life in cold water. In addition some species are capable of very deep diving. For example, Weddell seals have been known to plunge 1800 feet and to hold their breath underwater for an hour. Such activities require highly specialized physiological adaptations, but these will be discussed in the following section on whales, because it is in this group of mammals that the ability to live in water and yet breathe air has reached its highest expression.

Other mammals besides seals returned to the sea during the course of evolution, but of these the cetaceans—the whales, dolphins, and porpoises—are the most perfectly designed for moving and feeding below the waves. These highly specialized mammals completely cut their link with the land and evolved fishlike shapes that minimize resistance to their movement through water. Except for the very large heads of the sperm whale and right whales and the

The teeth of a killer whale (left) are very large conical pegs. It has 20 in each jaw, and when its mouth is shut these teeth interlock. Many of the smaller toothed whales have a similar arrangement, but the largest of the toothed whales, the sperm whale, has teeth only in its lower jaw (right).

thin fin on the back of some species, projections that break the elegant streamlining are kept to a minimum. Whales even lack a visible external ear; instead, the ear passage, filled with a wax plug, leads inward from a small slit in the surface of the skin. These adaptations for life in the water help to explain why people once believed that whales were fishes.

Nevertheless, a close examination shows that they are indeed mammals. Their skin is very smooth, for there are no scales. Although some species are apparently hairless, others have a few scattered bristles around the mouth, snout, and chin. Like all mammals, whales breathe atmospheric air, so there are no gills, and the nostrils open externally through a blowhole on top of the head—an important adaptation that allows a whale to breathe while floating horizontally at the surface. This blowhole leads to the lungs by a windpipe completely separate from the gullet, a modification that stops water entering the lungs from the throat.

Unlike seals, cetaceans have lost the hindlimbs of their terrestrial ancestors, but the pelvic girdle is still represented, although only by two small bones embedded in the muscle of the body wall and disconnected from the backbone. The forelimbs are also still present, but they have become modified into paddlelike flippers with four or five digits (depending on the species) buried in fibrous tissue and thus externally invisible. These foreflippers are freely movable only at the shoulder joint and are used primarily for steering and balance. Propulsion comes from the rear of the body and is provided by a flexible tail that bears two very large, horizontally flattened blades of tough fiber (the *flukes*) at its tip. The tail is powered by massive muscles that move the the whole assembly up and down (in contrast to fish in which the tail moves from side to side).

The tail works in such a way that the flukes always present an inclined surface to the water. This forces the water backward and so propels the animal forward. Ease of movement through the water is enhanced by the small degree of movement permitted the skin over the blubber layer underneath—a feature that reduces turbulence and streamlines the flow of water past the body. The power generated by some of the larger whales is immense and many are the stories of heavy catcher boats being dragged for considerable distances by large harpooned whales.

Perhaps an even greater indication of their power is the fact that these whales can drive themselves almost vertically clear of the surface of the water—an act known as "breaching." Why this is done is uncertain. Some scientists believe that whales breach in an effort to rid themselves of the many barnacles that are frequently fastened to them; others think that the resulting slap on the sea surface as they fall back serves as a sound signal; yet others suggest that whales breach merely for fun. But, whatever the reason, the sight of a whale perhaps 80 feet long, erupting from the surface of the sea, to land with a roar of cascading water, is spectacular.

Such a sight is uncommon, however, for whales breach only occasionally. A whale spends most of its life under the water and all that is usually seen is the top of its head as it comes up to breathe. As the blowhole breaks the surface the breath is expelled quickly and with some force in the form of a visible vaporous fountain known as a "blow." This seems to consist partly of condensed moisture (especially in cold polar regions), partly of foam from the windpipe, and partly of water expelled from a depression over the blowhole. The whale then breathes in, and the blowhole closes as the animal submerges. Even in large whales the whole process takes only two or three seconds and the animal may not show above the surface for more than about seven seconds. Some species—such as the rorquals, which seldom dive deeply—seem to blow a few times at intervals of 20 seconds or so and then dive for up to 10 minutes. But sperm whales, which habitually dive much deeper, blow 10 to 15 times at short intervals and then submerge or *sound* for 20 minutes or more.

There is no clear zoological distinction between the three groups that make up the order Cetacea. By popular usage, whales are those cetaceans longer than 15 feet when mature, porpoises reach about six feet, and dolphins six to 12 feet. Zoologists in fact divide living cetaceans on a different basis and into two suborders: the Odontoceti, or toothed whales, which possess teeth, and the Mysticeti, commonly known as the baleen or whalebone whales, in which teeth are completely absent after birth. The two groups also differ in their feeding habits, geographical distribution, and sexual behavior, and in the number and position of the blowholes.

The toothed whales include the sperm whale, the largest of the group (adult males can be 60 feet long), the white whale, the killer whale, the narwhal with its curious tusk, and the dolphins and porpoises. Most of these species usually eat squid, octopus, fish, and, occasionally, crustaceans. Their simple, often sharp, cone-shaped teeth are well-adapted for catching and holding their slippery prey, but are not differentiated into incisors, molars, and so on, for biting and chewing. The prey is usually swallowed whole, only the larger pieces being torn into smaller chunks. The numbers of teeth in each jaw vary considerably from species to species. In the sperm whale, for example, the lower jaw is armed with many well-developed teeth, but the upper jaw has only rudimentary ones buried in the gums. By contrast, the killer whale, which includes penguins, seals, and whales in its diet, has 20 large oval teeth in each jaw. These are set firmly in their sockets and are so arranged that those of the upper jaw interlock with those of the lower when the mouth closes. In a few species the teeth never erupt at all, and in others there are just one or two in each jaw.

The baleen whales owe their name to the fact that instead of teeth they possess numerous parallel plates of baleen that hang down from the gums of the upper jaw. This springy substance is the celebrated "whalebone" that, in the heyday of the whaling industry, was much in demand for the manufacture of goods ranging from coach whips to ladies' corsets. The name "whalebone" is a misnomer, for baleen bears no resemblance to bone and is, in fact, chemically almost identical to the hair and nails of man. For a short period before birth, young baleen whales bear rudimentary teeth, but these are quickly lost and replaced by triangular plates of baleen that grow downward from the palate. In the adult animal from 150 to 400 of these plates (the number depends on the species) hang down on each side of the jaw like a curtain. The plates are placed very close together and are much frayed on the inner edges, forming an effective matlike sieve.

The feeding habits of these whales are completely different from those of the toothed whales, for their diet consists almost entirely of krill, although they are opportunistic feeders and several species include an occasional fish on their menu. Their basic food chain is one of the shortest there is: from planktonic plants to planktonic animals to whales.

There are two main groups of baleen whales, and each has evolved to feed in a different way. One group consists of the right whales, so called because they were the "right" whales for whalers to catch, being relatively slow swimmers and floating when dead. Right whales employ a feeding method known as "skimming," whereby they merely swim through a mass of krill with the mouth open. Seawater constantly flows into the mouth and out through the baleen plates, but the krill remains behind caught in the mesh. To enable them to take in large quantities of food at one time, these whales have developed huge mouths, and long baleen plates enclose the space

The many longitudinal grooves extending along the throat and chest of the rorquals, seen here (right) on a fin whale, are believed to open out when the mouth is opened, making this cavity considerably larger. By contrast, the right whales do not have these furrows but, instead, have developed enormous bow-shaped heads with huge mouths. In these whales, the upper jaw is arched, but the lower jaw is w-shaped.

The baleen blades of a minke whale are clearly shown in this photograph of the roof of its mouth. Baleen consists of numbers of parallel plates of horny material with frayed inner edges. The individual blades of the rorquals are much shorter than those of the right whales. Feeding is accomplished by gulping large quantities of seawater and shrimplike krill, and pressing the water out through the baleen filters.

between the upper and lower jaw. For example, in the Greenland right whale or bowhead (the only Arctic representative of this group), about one third of the total body length of some 50 to 60 feet is occupied by the mouth. The upper jaw is arched and from each side hang more than 300 baleen plates, some of them 15 feet long. The lower jaw is U-shaped in front and raised high on either side, and the very large, well-developed lower lips lie close up against the plates. Such massive heads may give these whales a clumsy appearance but they certainly provide a very efficient scoop for krill and small fish.

By contrast, the rorquals, the largest single group of baleen whales, are much more streamlined because they lack the huge heads of the right whales. Instead, they possess a series of longitudinal folds or furrows on the throat and chest, which open out like a concertina when the jaws gape, vastly increasing the mouth's capacity. The upper jaw is not arched, so the baleen plates are relatively short. These whales feed by "gulping"—making an open-mouthed lunge in the direction of food, then closing the mouth and using the tongue to force the trapped water out through the baleen mesh. What is left behind is

then swallowed. It seems that rorquals are not fussy about their food, for their stomachs sometimes contain the remains of seabirds that have been gulped in with a mouthful of krill.

Rorquals include the largest of all whales. Indeed, the biggest animal ever to have lived is a member of this group. This is the blue whale or Sibbald's rorqual, also called the sulfur-bottom because of the yellowish color given to its undersurface, colored by the film of diatoms that frequently adheres to it. A blue whale may grow to be 110 feet long and to weigh 145 tons, equal to the combined bulk of about 1500 men or 24 elephants. In comparison, the largest known dinosaur was only 70 feet long and weighed a mere 36 tons or so. Water, with its natural buoyancy, is the only medium in which such an immense animal could live. On land the whale's sheer weight would crush it to death. Naturally, a creature of such proportions requires vast

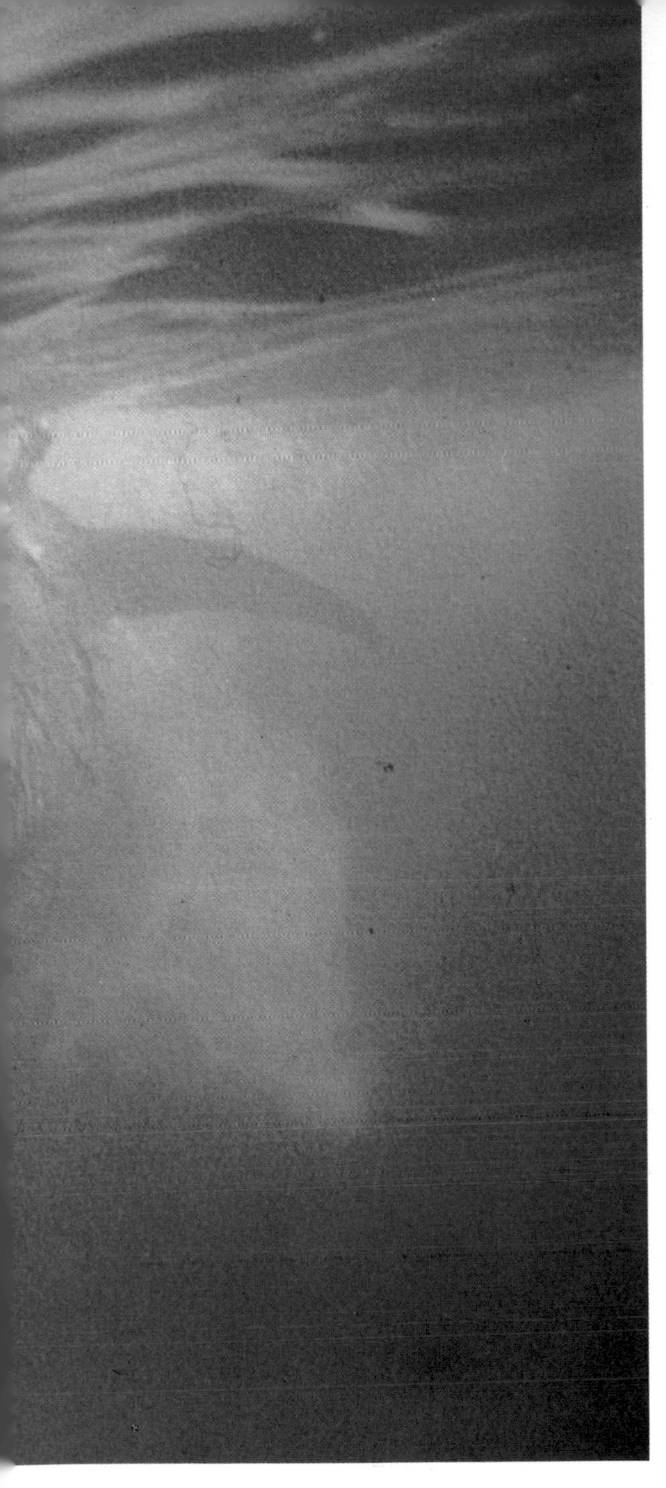

The streamlined shape of a rorqual is revealed in this photograph of a minke whale. Rorquals are among the swiftest animals in the seas: their average cruising speed is 10–12 mph, and they "sprint" at 18–20 mph. The sei whale approaches 30 mph.

amounts of food, and biologists estimate that a full-sized blue whale can consume between two and four tons of krill a day.

By way of contrast, the smallest rorqual is the minke (pronounced *minky*), or piked, whale or lesser rorqual, about 30 feet long. Other rorquals are between the blue and minke whales in length. The fin whale or common rorqual reaches about 80 feet, the sei (pronounced *sigh*) whale and Bryde's whale (primarily a temperate and subtropical species) attain about 50 feet, and the humpback whale averages 40 feet.

Because baleen whales feed near the surface, they need dive neither deeply nor for very long, although fin whales apparently descend at least 960 feet and can remain submerged for up to 30 minutes. On the other hand, toothed whales must often hunt for food at considerable depths, and sperm whales have been reliably recorded to descend to 3250 feet and to remain underwater for 90 minutes or more. A whale's body is largely composed of incompressible liquids, so the tremendous pressure (about 1600 pounds per square inch) exerted by the seawater at such depths does no damage to it. Moreover, whales can quickly surface after a prolonged deep dive without any ill effects. In humans, a rapid ascent following such a dive would result in caisson disease, or "the bends." This crippling, often fatal, condition is caused by the nitrogen in the air that a submerged diver continuously breathes. Nitrogen dissolves harmlessly in the bloodstream under pressure, but if the pressure is rapidly reduced (sudden decompression), it comes out of solution too quickly, forming gas bubbles in the blood vessels, which cause pain, paralysis, and death.

The ability of some marine mammals to make deep and prolonged dives impossible for land mammals has long been the subject of much study. Satisfactory answers to all the questions are still lacking, but at least a few of the general mechanisms involved are understood. For instance, it is known that marine mammals avoid painful decompression effects because they are not continuously supplied with air while underwater. Contrary to what might be expected, seals and whales dive with very little air, and consequently little nitrogen, stored in the lungs. In fact, seals breathe out before diving (thus achieving neutral buoyancy, which enables them to remain submerged with comparatively little effort), and before sounding whales carry less than 10 per cent of the necessary oxygen as air in the lungs. Moreover, their lungs collapse under pressure, forcing any unabsorbed air back into the relatively incompressible windpipe, from which no gas can enter the bloodstream. Some

The narwhal is a fascinating Arctic species that has only once been captured alive and put on display (at New York Aquarium in September 1969). Its name may have come from the Scandinavian word nar, *meaning corpse, for the skin has a color rather like that of a human corpse. The narwhal's most outstanding feature is its remarkable ivory tusk, found only in the male, where it may reach a length of 9 feet or more. This formidable-looking adornment to the male's 10-foot-long body is a tremendous enlargement of the left canine tooth, and always twists spirally to the left. Very occasionally, both upper canines form tusks. Their function is unknown, but they may be used to break through ice that is threatening to trap the animal. Other suggestions are that the tusk may be used in fighting or to stir up the seabed in the search for shrimps. On the other hand, it may be a secondary sexual characteristic. In medieval times the tusks were thought to be unicorns' horns, and were literally worth their weight in gold. They were thought to cure every known ailment, and were highly prized as detectors and neutralizers of poisons—a fanciful legend that made them seem a very valuable possession to noblemen of the time.*

experts think that the high pulse rate noticed in seals and whales following a return to the surface flushes out any small excess of nitrogen that may have found its way into the blood.

But if these mammals take down such a small amount of air, how do they get enough oxygen to last for up to 90 minutes without taking a breath, when in most cases man passes out after three minutes? There are a number of mechanisms at work here. First, volume for volume the blood of these mammals contains up to twice as many hemoglobin-containing cells as the blood of land mammals. Hemoglobin carries oxygen. Consequently a marine mammal can store up to twice as much oxygen as a land mammal of similar bulk. (Seals, in addition, have *more* blood in proportion to their volume than do their terrestrial counterparts.) Then, too, oxygen is stored in the muscles, which contain large quantities of *myoglobin*—an oxygen-carrying substance similar to hemoglobin. Thus these marine mammals have much higher concentrations of oxygen stored in their bodies at the beginning of a dive than terrestrial mammals have at any time. In addition, they utilize it much more efficiently. Underwater, the heartbeat slows considerably, dropping in whales by as much as 50 per cent and in seals from a normal 150 beats per minute or so on the surface to as few as 40 or even 10 beats per minute depending on the depth of the dive. Coupled with the fact that these animals can greatly restrict blood flow to all but vital organs such as the brain and the heart, this enables them to eke out their oxygen ration for up to an hour.

Furthermore, once the oxygen supply in the muscles is used up, these tissues continue to function for a short time by converting some of the carbohydrate present to lactic acid, thereby releasing energy (and carbon dixoide) without the use of oxygen—a process known as *anaerobic respiration*. This process incurs what is known as an "oxygen debt" that is repaid on surfacing when the lactic acid is broken down to carbon dioxide and water and the animal breathes deeply and quickly several times to flush out the carbon dioxide and build up the store of oxygen in the myoglobin again. A marine mammal is further helped by the fact that the breathing control center in its brain is relatively insensitive to accumulations of carbon dioxide in the blood—a feature that allows the creature to hold its breath a great deal longer than any terrestrial mammal.

Cetaceans are found in all oceans and seas. Some species apparently occur only in the Southern Hemisphere. Three such examples are the barely known Wilson's hourglass dolphin, which has been seen just to the north of the Antarctic pack ice, Burmeister's porpoise, and the spectacled porpoise. Other cetaceans, including the white-sided dolphin and the white-beaked dolphin, are found only in the Northern Hemisphere. The common, or harbor, porpoise is widely distributed in the North Atlantic and North Pacific oceans, extending into the Arctic in both regions. Other northern species include the white whale, which lives and breeds almost exclusively in Arctic waters, and the fascinating narwhal, whose chief claim to fame is the male's remarkable ivory tusk, which in medieval times was thought to be the horn of a unicorn and to have medicinal properties.

The strikingly patterned, black-and-white killer whale, found the world over, is common in Arctic regions and by no means uncommon in the Antarctic, where penguins, seals, and even other whales flee before its voracious appetite. The sperm whale, immortalized in Herman Melville's *Moby Dick*, is largely circumtropical and subtropical in distribution, rather than polar, but some males, probably those worsted in the vicious fights of the mating season and thus not engaged in breeding, find their way to the colder waters of the Arctic and Antarctic. Sperm whales do not appear to make definite migrations, probably because their food is readily obtainable in most areas, but there is some evidence that the main breeding herds move northward in the northern summer and southward in the southern summer.

As a group, the rorquals perform the most impressive migrations of all cetaceans. The krill on which they feed is most abundant in polar areas, so the summer months are spent in feeding near the Arctic or Antarctic pack ice, but toward winter the whales begin a long trek to their breeding grounds in subtropical waters. There, after a pregnancy of about one year, the cows give birth, usually to a single calf, which probably lacks sufficient blubber to insulate itself against the cold polar waters. The rigid life cycle of these whales places a heavy physiological burden on the mother. She produces a calf weighing as much as 2 to 4 tons (an incredible weight when one considers the relatively short gestation period); she must then produce enough milk to enable the baby to grow considerably and to be ready for weaning a bare six or seven months later, while at the same time living chiefly off her blubber reserves. Mother and calf then travel back to polar waters, where the cow has only a short resting and feeding period before returning to the breeding grounds again to mate. The reproductive cycle in these whales is thus one calf every two years. In this they differ from the sperm whales, whose normal cycle is only one calf every four years, pregnancy lasting for about 15 months and the calf being suckled for a further two years.

From what has been observed in smaller cetaceans, it appears that cows giving birth are often assisted by other females acting as "midwives," particularly in the important task of nudging the newborn calf to the surface for its first breath. Such assistance is not confined to calves, however, for there are several recorded cases of whales coming to the aid of an injured adult and supporting it at the surface. Such rescuing activities imply a degree of social behavior and—so it has been argued—intelligence. Certainly cetaceans, particularly dolphins, have large brains—larger in absolute terms than those of humans—and species kept in captivity have responded quickly and easily to training. This has helped to show that except for the sense of smell (which in baleen whales appears to be slight, and in toothed whales nonexistent), the cetaceans' senses, especially that of hearing, are well-developed. But assessing the actual intelligence of mammals constructed so unlike ourselves remains a difficult problem. Possibly all that can be said at present is that dolphins seem to display an intelligence equivalent at least to that of the higher primates such as gorillas and chimpanzees.

The fact that individual cetaceans cooperate implies that they can communicate with each other. It has been long known that whales produce a variety of noises such as whistles,

Although whales apparently have a highly efficient method of locating obstacles in the water, they occasionally run ashore, become stranded, and die, as has the sperm whale (right). Why they do this is not known. Recent research has shown that many of these stranded whales have been suffering from ear parasites that may have affected their hearing, but whether or not this is the reason has yet to be proved.

Below: dolphins performing at the Miami Seaquarium, Florida. Much scientific research has been done on the intelligence of these graceful marine mammals. Dolphins, and indeed cetaceans in general, have relatively large brains and respond quickly and easily to training. Their senses, except for that of smell, are well developed, and they can communicate with each other and with their trainer by producing sound.

clicks, and groans. Indeed, the humpback whale can fairly be said to sing, producing a sequence of moans and whistles at varying frequencies to form a "song" up to 10 minutes long, repeated over and over. Each whale appears to have its own tune, and it may well be that these "songs" enable the animals to locate and identify one another over considerable distances. It is not known precisely how whales make sounds, but in general terms they do so partly by means of air sacs within the blowhole and partly by the passage of air through the larynx.

Besides making sounds to find one another, toothed whales appear to use very-high-frequency sounds for detecting food and obstacles in the water, emitting bursts of from five clicks up to several hundred clicks per second. This echolocation technique is similar to that used by bats in the air and perhaps by some submerged seals. The echo bounced back from an object appears to be received through the whale's lower jaw and possibly the sides of its forehead, whence the resulting vibrations reach the ear.

With such a refined technique for locating obstacles, it seems strange that whales occasionally run ashore, become stranded, and die on the beach. What is even more mystifying is the fact that a stranded whale, rescued and hauled back into the ocean before it is too late, frequently turns around and runs ashore again. Recent studies indicate that many stranded whales suffer from ear parasites, and these may impair the whales' hearing, and hence its echolocation capability, but this theory is not fully proven. Why these animals become stranded is just one of the many as yet unsolved mysteries of whales.

Man and the Polar Regions

Man has inhabited the Arctic from time to time for at least the last 30,000 years, since Old Stone Age peoples moved northward during a warm phase in the last ice age. At this time grasslands and forests extended northward to the Arctic coasts, and man could find food there both on land and along the intertidal zone. As the northlands entered the most recent spell of glaciation (known as the Würm glaciation in Europe and the Wisconsin glaciation in North America) these early peoples retreated south, and enormous areas were covered by sterile ice sheets. At last the climate grew milder, the ice sheets receded, and by 10,000 years ago people from many points in North America, Europe, and Asia were recolonizing lands newly emerged from the ice. Their numbers were small, and the land was still harsh. Wandering groups and relatively stable settlements must often have been hard hit by winter storms, or wiped out by starvation, but pioneers such as the Eskimos and Lapps gradually adapted to their cold surroundings.

Meanwhile, no one had ever seen or set foot upon Antarctica. Proof that there was land in southern polar regions arrived a little less than 400 years ago. True, philosophers in antiquity had surmised that it was there; to their tidy minds it seemed obvious that there must be a big landmass in the south to balance the known ones in the north. Because the nature of this continent was totally unknown, imaginations could run riot—and they did! The northern shores of this fabulous land were believed to lie somewhere just south of the equator, and it supposedly extended, warm and lush, to the South Pole. Slowly, this mythical land dwindled in size, pushed farther and farther south by successive voyages of exploration. It was not in the tropics, it was not in the temperate regions; then, at last, in 1599, the crew of a storm-tossed Dutch ship glimpsed icy mountains south of the southern tip of South America. They were probably the peaks of the South Shetland Islands, and not part of the Antarctic continent; nevertheless they indicated that whatever land lay still farther south was unlikely to be suitable for colonization or trade.

Throughout the 18th century, the voyages of

Members of Wally Herbert's British Trans-Arctic Expedition of 1968–9 wore furs and took provisions on sledges hauled by huskies. By thus traveling Eskimo fashion, they successfully crossed the entire Arctic Ocean in a nearly 3000-mile trek from Alaska to Spitsbergen. This transarctic journey via the North Pole ended an era of polar reconnaissance that had begun in the Stone Age.

Equinoctial
Amerique
Bresil
Peru
Riuiere de plate
Tropicque de capricorne
Terre Australle

exploration followed each other, finding remote islands in the Southern Ocean: Bouvet Island in 1728, Marion Island, the Crozet Islands, and the Kerguelen Islands in 1772, and South Georgia in 1775. James Cook, who discovered South Georgia (and subsequently the South Sandwich Islands), was not impressed. He wrote of it: "Thick fogs, Snow storms, Intense Cold and every other thing that can render Navigation dangerous one has to encounter and these difficulties are greatly heightned by the enexpressable horrid aspect of the Country, a Country doomed by Nature never once to feel the warmth of the Suns rays, but to lie for ever buried under everlasting snow and ice."

Despite these gloomy observations, and the suggestion that: "The Ports which may be on the Coast are in a manner wholy filled up with frozen Snow of vast thickness, but if any should so far be open as to admit a ship in, it is even dangerous to go in, for she runs a risk of being fixed there for ever, or coming out in an ice island," there were those who were not deterred. Cook's voyages were responsible for opening the floodgates of exploitation in Antarctic and sub-antarctic seas, for it was the publication of journals resulting from his explorations that also told the world about the huge populations of seals and penguins living there. Unfortunately, the news arrived just at the right time for commerce, because the traditional sealing grounds in the Northern Hemisphere had become virtually exhausted from ruthless and uncontrolled over-hunting. So, a mere three years after Cook's discoveries, American and European sealers began operations on islands off the coasts of southern South America, Africa, Australia, New Zealand, and, a little later, on South Georgia.

Now, though, let us return to the Arctic, the place that attracted most of the early polar explorers. Although men have inhabited the Arctic intermittently from about 30,000 years ago, the first civilized West European known to have journeyed there, about 320 B.C., seems to have been Pytheas of Massalia (now Marseilles). This Greek explorer wrote a book, or perhaps two books, about his travels. But his writings have long been lost. The Greek geographer Strabo paraphrased them, about the beginning of the Christian era, however, apparently trying to show that Pytheas had been a liar. Be that as it may, Pytheas certainly made a long sea voyage to a country he called Thule, which was possibly Iceland, but may have been northern Norway. He reported that it was one day's sail "from the frozen sea," and described a phenomenon he called the "sea lung." His description could well fit the slow rise and fall of pack ice under the influence of the oceanic swell (a motion that resembles a slow breathing of the sea), or even the mist often seen at the edge of sea ice, and recalling the visible breath of a man on a cold day. But we shall never know precisely what Pytheas meant.

World map made for an atlas of 1555 by the French cartographer Guillaume le Testu. It shows a hollow globe cut in halves printed on both sides. Although Antarctica was undiscovered and the Arctic mainly unexplored, le Testu rightly guessed that the North Pole stood in an ocean and the South Pole on a landmass.

There were seemingly no similar voyages to the Arctic for more than 1000 years, until Irish hermits reached Iceland in open boats in 790 A.D. But the chief pioneers of Arctic exploration were the Norsemen, who began colonizing Iceland from Norway in 875 and probably sighted Greenland only two years later. In 982 Erik the Red arrived in Greenland and later probed the west coast as far north as the site of modern Godthaab. During the next two years he probed even farther north before returning to Iceland. For centuries, until the climate grew intolerably harsh, farmers and fishermen retained a foothold on the coasts of Greenland, and by the year 1000 this polar island had become a springboard for the discovery of North America, but Norse discoveries in the Arctic seemingly remained unknown to most Europeans.

Exploration based in Western Europe turned its face to other areas, and culminated in the 15th and early 16th centuries in the discovery of sea routes to eastern Asia and to America. These finds in turn sparked off a search for new and shorter routes to the rich lands of east and southeast Asia—a search that took explorers deep into the Arctic. The year 1553 saw the start of the first voyage to seek a Northeast Passage—a short cut to China by way of the Arctic seas north of Eurasia. Sir Hugh Willoughby, an English soldier, commanded this expedition, with Richard Chancellor as navigator and second-in-command. Chancellor, with the strongest ship, made his way into the then unknown White Sea north of Russia, landed near Archangel, and journeyed overland to

William Hodges' watercolor shows James Cook's ships Adventure *(in the foreground) and* Resolution *at 61° South, collecting sea ice as a source of fresh water. Early in 1773 the ships became the first to sail across the Antarctic Circle. Cook's voyage proved that the Southern Ocean was inhospitable, but his account of its seal herds sparked off a profitable, though brutal, sealing industry.*

Moscow. Meanwhile Willoughby, who had lost contact with Chancellor, pressed on eastward across the Barents Sea (a sea north of the White Sea) until he came to the more southerly of the two islands collectively called Novaya Zemlya. Here he turned north and made slow progress against high winds and coastal ice. Seeking winter quarters, he turned south to the Kola Peninsula of mainland Russia. At that time no Englishman had had experience of the hazards of an Arctic winter, and Willoughby and his crew all died.

Willoughby's fate made plain that there was no easy northeast route to the Orient from Europe. Some explorers began instead to seek a Northwest Passage to China by way of Arctic North America. In 1576 the English navigator Sir Martin Frobisher set sail from Deptford, near London. Arriving at Baffin Island, to the west of Greenland, he found some "black rock" that he thought contained gold (it eventually turned out to be specks of mica in quartz). Frobisher loaded his ship with the worthless stuff and returned home. Although he made two more expeditions in the same direction, he was now obsessed by the "gold" and made no more serious attempts on the Northwest Passage. Others did, however, among them John Davis in the 1580s and Henry Hudson at the beginning of the 17th century. Both probed straits lying west of Greenland. The five voyages of the great Arctic explorer William Baffin took place between 1612 and 1622 and revealed much of the geography of the northeast Canadian Arctic.

Hudson also reinvestigated the possibilities of a Northeast Passage, following in the footsteps of the Dutchman Olivier Brunel, who had reached Novaya Zemlya in 1584, and Willem Barents, who had broken through sea ice east of Novaya Zemlya and entered the Kara Sea in 1594. Barents was a skillful sailor, but he was also very lucky, for 1594 was a "good" ice year, with unusually mild conditions. His feat was not repeated for more than 200 years. Ice conditions were severe for his second expedition, and he could not get beyond Novaya Zemlya, but on his third voyage he discovered Bear Island and Spitsbergen, both north of Norway, before being shipwrecked off Novaya Zemlya, where his expedition became the first to survive the full rigors of winter deep inside the Arctic.

By the mid-1600s Russian sailors were investigating the seas off the Siberian coast. In 1648 Simon Dezhnev traveled from the Kolyma River to the Anadyr River, a journey that would have taken him through the Bering Strait and right around the extreme northeastern corner of Siberia if —as Dezhnev claimed—he went by sea (and not by land, as some believe). Credit for discovering the Bering Strait that separates Asia from North America usually goes to the Danish navigator Vitus Bering who undoubtedly traversed the strait in 1728 in Russian service. Only persistent mist prevented him from seeing Asia to the west and North America to the east, and thus guessing that no land bridge connected them. Bering was subsequently put in charge of Russia's Great Northern Expedition, which mapped almost all the northern Siberian coast between 1734 and 1742. Even so, no one could be sure whether Asia and North America were separate until the Norwegian explorer Nils Nordenskjöld sailed in the 300-ton ship *Vega* from the north Norwegian port of Tromsø to the Pacific via the Arctic Ocean in 1878–9. His expedition was the first to complete the Northeast Passage.

Meanwhile, exploration of the North American Arctic was more piecemeal, but by 1860 some 7000 miles of coastline had been seen by relief parties seeking Sir John Franklin's expedition, which had disappeared in 1848. In 1906 the Norwegian explorer Roald Amundsen accomplished the first unbroken journey through the Northwest Passage.

In the 20th century, men have probed both polar regions to the very poles. In 1909 the American Robert E. Peary reached the North Pole, and in 1911 Amundsen hoisted the Norwegian flag at the opposite end of the earth. Since then, sophisticated expeditions have safely spent long periods deep inside the polar regions. In Amundsen's day, however, polar exploration was still very dangerous. Only three months after his success, his British rival Robert Falcon Scott died blizzard-bound in the Antarctic, a victim of bad luck and bad planning like so many polar explorers before him.

In retrospect, it seems incredible that for centuries little was done to equip Europeans properly for long polar journeys. Drawings and paintings show them in clothes that would have been barely suitable for winter in London or New York. A stout, thick, woollen cloth called "Fearnought" was introduced in 1772 to make coats for the seamen, but that seems to have been the limit of special clothing. And yet the explorers had the warmly clad Eskimos and other Arctic peoples to learn from.

Men from temperate regions are not used to great extremes of temperature, but laboratory experiments show that animals can become physiologically acclimatized to cold. Therefore one might think that acclimatization could also occur in man, provided conditions were right. To examine this notion, let us begin by seeing how the body keeps warm in normal conditions and how it reacts to chilling.

Like all warm-blooded animals, man has a highly efficient mechanism for keeping his temperature constant in spite of variations in the environmental temperature and the changing amounts of heat produced in the body. Although body temperature is higher by day than by night, the variation is seldom more than about 1°F. The average temperature of a healthy person is usually 98.4°F.

To say that the average body temperature is 98.4°F does not, of course, mean that the whole body has to be this warm to remain healthy or comfortable. In terms of temperature requirement we can think of the body as consisting of two parts: an inner "core" normally maintained at 98.4°F; and an outer "shell" that feels comfortable at 91°F but may vary quite considerably in temperature without any ill effect. Shell temperature slowly falls if the body is continuously exposed to cold, the rate of fall depending both on the outside temperature and on the wind-chill factor. But shell temperature can drop to 32°F before the outer tissues start to freeze and to turn white and numb—the medical condition called *frostbite*. Frostbite results in the loss of toes and fingers if untreated. But even if his shell temperature drops from 91°F by more than 50°F a man will survive provided his core temperature stays high. Unless the environmental temperature is very low indeed, a man can maintain his core temperature by keeping active enough to generate adequate internal heat. If he gets so tired that he has to rest, however, the amount of heat produced internally will drop and gradually his core temperature will fall. This may not happen to any marked extent for many hours, and until it does he can make good the temperature loss by taking exercise. But once core temperature falls, it becomes difficult to resume activity, and so a vicious circle is set up. Semiconsciousness sets in when core temperature falls to 91°F, and the victim loses consciousness when it drops to 86°F. Expert treatment can still effect recovery at this stage, but the chances of survival become remote once the temperature falls to about 77°F, although there is no precise lethal temperature.

Maintaining body temperature in polar climates, then, depends on keeping a careful balance between heat loss and heat production. Heat loss can be reduced by the insulation provided by adequate clothing. Present-day polar clothing gives excellent protection, being both warm and windproof. Heat production occurs inside the body when the sugars, fats, and sometimes proteins, in food break down in the presence of oxygen, yielding heat and the waste products carbon dioxide and water. Periods of strenuous activity, such as sledging, will require more energy and thus a greater intake of energy-rich foods than relatively inactive periods in a base camp. Of course, like anyone else, people living in the polar regions also need adequate supplies of proteins and the minerals, vitamins, and water without which no diet is complete.

But although physically active polar explorers may eat more than most men, they do not seem to convert food into heat energy any more efficiently. Thus, man's physiological ability to adapt to cold does not seem strongly marked on a short-term basis (although on a long-term basis the Eskimos have evolved short, stocky bodies that lose heat less readily than tall longlimbed ones). However, an individual's body weight and the thickness of insulating fat deposits beneath the skin do tend to increase. Psychological adaptation to cold may be indicated by the fact that, after men have spent a time at sub-zero temperatures, they grumble about a "heat wave" if the temperature rises to freezing point.

Polar cold makes an immediate impact, but man also suffers from the more insidious effects of the long days and nights experienced in the far north and the far south. People forced to endure dark, polar winters reportedly suffer from depression and commit suicide more readily than people wintering in lower latitudes. Moreover, many visitors to polar regions blame the long summer day for difficulties in sleeping. Although there is no proof that long spells of light and darkness affect the normal daily rhythms of every human body, research shows that at least some individuals unaccustomed to such conditions suffer a disrupted sleep pattern. Physiologists have found that the average amount of sleep required in polar regions is about the same as in temperate latitudes—about eight hours in every 24. But during the polar winter some people may take a number of short naps in each 24-hour period instead of one long sleep. When spring comes, they resume a normal sleep pattern, before prolonged summer daylight causes another, though milder, disruption.

Physical and emotional stress, such as polar explorers experience, may lead to a number of physiological responses, including a rise in the amount of adrenaline produced by the body's suprarenal glands. Besides increasing the blood flow to brain and muscles, this change in adrenaline flow reduces the numbers of *eosinophils*—certain white blood cells—in circulation. Physiologists have noticed this effect of stress among sledge-party leaders faced with difficult decisions, although the physically arduous work of sledging appears to leave the number of eosinophils unaltered. Surprisingly, perhaps, it seems that being duty cook on a polar research team is decidedly stressful, for scientists find that there is a marked drop in the eosinophil count of the men taking their turn at this job, particularly toward the end of their spell!

We have seen how man discovered the ends of the earth and how those places affect the people who live there. Now let us turn to man's impact upon the wildlife of these regions. We have already mentioned that Captain Cook's discoveries led to the exploitation of Antarctica's seals and penguins. Seals were to bear the brunt of man's early onslaught upon the fauna of the far south. At first the fur seals were the main target, because there was a ready and lucrative market for their pelts in Europe, North America, and China. In 1790 two ships obtained cargoes of furs from South Georgia, and from that time the herds there suffered heavy attacks. By 1822, Weddell estimated that at least 1,200,000 furs had been taken from South Georgia alone, and its

Right: the fur-clad figure of Robert E. Peary, who wore sensible polar clothing for his successful attack on the North Pole in 1909. (The American explorer is seen on the wooden steamship Roosevelt *that took him to Ellesmere Island from where his sea-ice journey began.) Below: ill-clad American sailors trying to keep warm on an Antarctic island in 1840, a painting based on a sketch by their expedition leader, the US naval officer Charles Wilkes. Wilkes complained bitterly that the clothing his government had provided was inadequate. Like countless other early explorers, his crewmen risked frostbite and even death by venturing into realms of intense cold without adequate means of conserving essential body heat.*

stocks were gravely depleted. But meanwhile, in 1819, in the South Shetland Islands, Captain Smith of the brig *Williams* had discovered the largest fur-seal herds in the entire Southern Hemisphere. The slaughter continued so effectively that within 10 years not a fur seal was to be seen among these islands. This was the beginning of the end for the first phase of subantarctic fur sealing, for the other subantarctic islands had only small groups of the creatures. Nevertheless, much money was tied up in ships and equipment, and shipowners refused to leave it idle; instead they started hunting elephant seals for their oil. And so South Georgia, the Kerguelen Islands, and Heard Island were once again scenes of carnage as sailors clubbed and shot elephant seals, before stripping off the blubber and boiling it in try-pots to extract the oil. They simply left the rest of each carcass to rot. Because the hunters slaughtered bulls, cows, and calves indiscriminately, they wiped out whole populations. Oil proved less valuable than furs had been, however, and the oil sealers abandoned their hunt at an earlier stage than the fur sealers had done, leaving small pockets of seals here and there.

About half a century after the slaughter at the South Shetland Islands, their surviving fur-seal herds had multiplied sufficiently to attract fresh sealing enterprises. Again, as many furs as possible were taken (probably about 45,000 between 1872 and 1888) until once more hunting became unprofitable.

By the beginning of the 20th century, the elephant-seal population on South Georgia was beginning to recover, and the Falkland Islands government, which controlled South Georgia, set out to regulate the hunting there. The basis of controlled exploitation was simple: although these seals were harem breeders, they produced almost equal numbers of males and females, and surplus males could be killed without upsetting the breeding potential of an entire herd. The Falkland Islands government licensed seal killing so that only males might be taken. It also divided the coastline of South Georgia into four

Man's impact on Arctic seal populations has increased over the centuries. Old-style Eskimo hunters, armed with harpoons, hunting alone, and killing for food (above), had no big effect on seal numbers. Europeans, hunting in groups, and committing mass slaughter for commercial reward (left), placed entire species at risk. Eskimos armed with rifles (right) now also kill sometimes for profit. Strictly enforced government quotas are the only way to ensure that harp and hooded seals will survive.

parts, one to be left unhunted each season. Also, the actual hunting season was restricted. As a result, between 1910 and 1964 (when the fishery ceased) sealers killed some 250,000 seals yielding a harvest of 74,000 gallons of oil, yet the numbers of live elephant seals had increased substantially.

There is every reason to believe that the Kerguelen fur seals have been gradually re-establishing themselves in their old homes, but it seems unlikely that either industry will be restarted in the foreseeable future. Both the Kerguelen fur seal and the elephant seal are probably safe for the time being, partly because economic circumstances make hunting these particular mammals unprofitable.

Disappointingly, man still mishandles one of the major biological resources of the south polar regions: he is overhunting the whales. But the whaling industry's excesses are being corrected by economic forces, to its disadvantage, for so few whales remain that it is no longer worth sending large fleets of factory ships and catchers to the Antarctic. Fear of hunting some species

into extinction drove the industry to fix quotas through its controlling body, the International Whaling Commission, which went further and introduced the international observer scheme in 1972. Under the scheme, observers from one country sail with the fleet of another, so as to monitor the catch. Next, in 1973, use of the iniquitous "blue whale unit" was ended. For many years whalers had employed this device to measure the quotas (one unit equalled one blue whale, two fin whales, six sei whales, or—before they were protected—two and a half humpback whales). Catching whales on this basis meant that as the blue whale became scarce, whalers turned their attention to the somewhat smaller fin whale, then to the still smaller sei whale. Despite an improved rationing system, the fast declining sperm and sei whales still bear the brunt of the whalers' attack, and even smaller species, including dolphins, are now being hunted.

So far we have described man's impact on animals in the south polar regions. But the Arctic, too, has seen much, often wanton, killing of sea mammals, especially seals. Sealing began in the Antarctic purely as commercial exploitation. In the Arctic, though, it began as a requirement of survival, and unlike sealers in the Southern Hemisphere, the Eskimos and other early hunters wasted nothing. They used all parts and by-products of the seals: for instance, meat for food, skin for clothing, and oil for heat and light. In general, the native Arctic people hunted singly, with primitive weapons, but different tribes used different methods. For example, in North America, the Mackenzie Bay Eskimos crawled up within range of any seals they could see and took them by surprise, whereas Victoria Island hunters sat by unoccupied breathing holes for hours waiting for the seals to appear. Under such conditions there was no risk of overexploitation. But the situation changed sharply when Europeans appeared in numbers and with firearms.

Several northern seals became heavily persecuted. One was the walrus. Commercial hunting by Europeans for its tusks, hides, and oil began in the 16th century, and by the 17th century walrus numbers were falling. By the second half of the 19th century hunters were gathering about 12,000 pounds in weight of walrus ivory every year—a rate the walrus population could not sustain. This marine mammal was altogether wiped out in parts of its range. The nations involved, however, eventually acted to save it. In 1931 Canada restricted walrus hunting in the Canadian Arctic to Eskimos and certain white residents, and forbade the export of raw skins and ivory. In 1949 such controls were intensified. From 1956 the USSR allowed only the Eskimo and Chukchi peoples to hunt the species, and the Danes and Norwegians now accord it total protection. Today, though, the Atlantic race is considered to be in danger, and although the Pacific race seems to be safe, its situation needs careful watching.

There were about two and a half million Pribilof fur seals when their breeding grounds in the Bering Sea's Pribilof Islands were discovered in the late 1780s. By 1868, when the United States bought the islands from Russia, Russian sealers had greatly reduced the seals' numbers. American hunters proved even more greedy, and by 1911 only about 200,000 animals remained. International measures to control hunting followed, and the population has now increased to about one and a half million.

Among the northern true seals the harp seal and the hooded seal have been, and still are, heavily hunted. In both species the pup is the primary target for its thick and attractive fur, and it seems likely that both seal populations have dropped despite the introduction of quotas. But the quotas for harp seals have been almost as high as the output of young in some places. Canada, however, has protected the harp-seal herds in the Gulf of Saint Lawrence from aerial hunting since 1970, and from hunting by large ships since 1972. Since 1965 Canada has also granted total protection to the small numbers of hooded seals in the Gulf.

Another outstanding Arctic sea mammal to suffer has been the bowhead, or Greenland right whale, which is now very rare. The native peoples of the Arctic hunted the whale from prehistoric times and it formed the basis of an important industry between 1611 and 1912. The bowhead has been free from commercial hunting since 1912, and protected by international convention since 1935, but its numbers appear to have made no real recovery.

Among Arctic land mammals, the polar bear has been the most heavily threatened by man. The decline of the polar bear dates from the 17th century, when excessive hunting accompanied the opening up of the Arctic. Its numbers progressively fell with the introduction of precision weapons, powerboats, and aircraft; and disease introduced by domestic animals may also have taken its toll. In 1973–4, however, the five nations with lands in the Arctic (Canada, Denmark, Norway, the USA, and the USSR) signed an agreement on the conservation of polar bears. The new accord should give almost complete protection to the species, because it forbids the capture or killing of polar bears except by local people using traditional methods, or for purposes of conservation and scientific research. This is the first treaty between the five so-called Arctic states, and wildlife enthusiasts hope it will lead to more joint conservation activities.

While piecemeal conservation was emerging in the Arctic, more sweeping measures have been brought into force in the south polar regions. Some measures appeared because the gradual introduction of controls on northern sealing made conservationists fear that the sealers might turn their attention to hitherto untouched stocks of Antarctic seals. For the first time in history, however, conservation legislation has been ahead of exploitation, for in February 1972 the Conference on the Conservation of Antarctic Seals met in London to decide on measures and limits in case a new sealing industry did emerge in the far south. The result of the conference was a convention giving total protection to the southern elephant seal, Ross seal, and the Kerguelen, South African, and South American fur seals. In order to protect Weddell seal breeding stocks at the time when these are most highly concentrated and vulnerable, the convention forbids the killing or capture of any Weddell seal one year old or more between 1 September and 31 January. The convention also sets quotas for crab-eater, leopard, and Weddell seals, prescribes closed seasons, and defines sealing zones and reserves.

The whole point of the convention is that it regulates sealing on the high seas south of latitude 60° South, and on floating ice in the region. Measures for conserving seals around the Antarctic continent, on islands south of 60° South, and on the ice shelves and other fast ice—but expressly excluding the high seas—had been set out under the provisions of the Antarctic

Slaughtered sperm whales tied by the tails to the side of a whaling vessel. Fast whaling steamships, heavy harpoons fired from cannons, spotter planes, and electronic detectors combined to make whaling so efficient that in 1964 alone 63,000 whales died to supply oil for margarine and meat for man and his pets. But such overhunting had brought some species close to extinction.

Treaty already. This was signed on 1 December, 1959, by 12 nations (Argentina, Australia, Belgium, Chile, France, Japan, New Zealand, Norway, South Africa, the United Kingdom, the USA, and the USSR) and the treaty came into force in June 1961.

The first sentence of the first article of the treaty states: "Antarctica shall be used for peaceful purposes only." The treaty goes on to deal with specific dos and don'ts, four of them vital for preserving the region's wildlife. These proclaim that native species of animals and plants shall not be unnecessarily disturbed, injured, or destroyed; alien plants and animals shall not be introduced; protected areas shall be declared to safeguard valuable ecosystems, plants, or animals; and individual nations shall be responsible for the behavior of tourists on the Antarctic lands they administer.

In Antarctica, conservationists have been able to plan to maintain the environment in an almost pristine state right from the start. It is the only continent where conservation guidelines were applied almost before man had begun upsetting the balance of natural plant-animal systems. Man is indeed still a relative stranger here, for Antarctica's human population is only about 500 in winter, and less than 5000 in summer. Twelve countries maintain a variable number of scientific stations, but territorial claims have been set aside for a period of 30 years from the date the treaty came into force. This last fact has encouraged suggestions that the entire continent should be declared an international park. For example, Recommendation 5 of the Second World Conference on National Parks (held in Yellowstone National Park in 1972) dealt with the subject, and the International Union for Conservation of Nature and Natural Resources has since brought this recommendation to the attention both of the governments concerned with Antarctica, and of the Treaty Organization. No doubt the political and legal problems in creating an International Antarctic Park would be great, but they are surely far from insoluble.

Apart from the whales, then, animal and plant life in the Antarctic seems to be in a satisfactory state. Among the wild inhabitants of the Arctic the numbers of many animal species have been grossly depleted, but few species appear at risk of extinction. Whether things will remain as they are is problematical, because conservationists increasingly clash with industrial interests, notably in the North American Arctic. Ecologists determined to safeguard the earth's rich but diminishing diversity of plant and animal species have urged the Canadian government to establish an enormous protected area in the Yukon, continuous with the immense (almost nine-million-acre) Arctic National Wildlife Range established by the USA in 1960. The combined reserves would then be called the Arctic International Wildlife Range. Such proposals for the blanket protection of vast tracts of land are largely aimed at saving them from the undesirable effects of mineral exploitation in the Arctic.

The prospect of 20th-century technology invading magnificent, harsh, and lonely wilderness was largely responsible for the widespread opposition of the early 1970s to the so-called Alaska Pipeline, planned to carry oil from wells in north Alaska to an ice-free port in the south. Polar ecosystems are much more fragile than those of temperate regions. Their plant and animal communities develop slowly because of adverse conditions, and survive only because of specific adaptations. These adaptations, however, render the species vulnerable to any disturbance. Ecologists have been particularly worried about how the pipeline's construction, presence, and possible leakages might affect caribou migration, breeding populations of waterfowl and birds of prey, rivers where fish spawn, and above all the tundra vegetation growing precariously on the thin soil above the permafrost. Some research has been done and protective measures have been written into the contracts. But the success of these measures will depend largely on the individual workers.

Man in polar regions has an unenviable record of destruction, greed, and thoughtlessness. Even the tourists now reaching the far north and far south in increasing numbers have irreparably damaged vegetation and colonies of breeding birds. Someone has remarked — only half jokingly—that parts of the Southern Ocean are paved with discarded cans thrown overboard from ships. Nevertheless, the future is hopeful because more people than ever before are taking an interest in the welfare of these remote areas, the last truly wild places on earth.

Worker preparing to add a new section to the Alaskan pipeline designed to link oil fields in north Alaska with the ice-free port of Valdez in the south. Environmentalists saw the pipe as a threat to Arctic wildlife and opposed it, but in 1973 fear of an oil shortage persuaded Congress to let building begin.

Index

Page numbers in *italics* refer to illustrations or captions to illustrations

Picture Credits

Key to position of picture on page: (B) bottom, (C) center, (L) left, (R) right, (T) top; hence (BR) bottom right, (CL) center left, etc.

Cover: W. Curtsinger/Photo Researchers Inc.
Title Page: George Holton/Photo Researchers Inc.
Contents Page: Nicholas DeVore III/Bruce Coleman Inc.
9 Charles Swithinbank
12 Andrew Clarke/British Antarctic Survey
13 Russ Kinne/Photo Researchers Inc.
14 Suinot/Explorer
15(R) R. Mickleburgh/Ardea, London
16 Suinot/Explorer
17(R) Andrew Clarke/British Antarctic Survey
19 W. Schmidt/ZEFA
21 Bill Ruth/Bruce Coleman Inc.
22(L) H. Groendal/ZEFA
23(T) Bill Ruth/Bruce Coleman Inc.
23(BL) Inigo Everson/B. Coleman Ltd.
23(BR) Ian Collinge/British Antarctic Survey
24 Charles Swithinbank
27 E. Hummel/ZEFA
28, 29(T) Bill Ruth/Bruce Coleman Inc.
29(BR) C. J. Ott/Bruce Coleman Ltd.
30 Francisco Erize/B. Coleman Inc.
31(B) Fred Bruemmer
32(T) Brian Hawkes/N.H.P.A.
32(B) B. L. Sage/Ardea, London
33(T) Dr. K. Bonath/ZEFA
33(B) Carré/Jacana
34(L) Jen&Des Bartlett/B. Coleman Inc.
34(R) Stephen Dalton/Photo Researchers Inc.
35(L) Brian Hawkes/N.H.P.A.
35(R) West/Frank W. Lane
37 Francisco Erize/B. Coleman Ltd.
38(T) G. Maxwell/British Antarctic Survey
38(BL) Inigo Everson/B. Coleman Ltd.
39 Payne/British Antarctic Survey
40 Guy Mannering/B. Coleman Inc.
41 M. G. Richardson/British Antarctic Survey
43 Stoll/Jacana
44 Photo Bryan Alexander
45(R) Joseph Rychetnik/Photo Researchers Inc.
47(T) Massart/Jacana
47(B) K. W. Fink/Bruce Coleman Inc.
48 Wildlife Unlimited/Photo Researchers Inc.
49 Varin/Jacana
52 C. J. Ott/Bruce Coleman Ltd.
53(T) Fred Bruemmer
53(B) Francisco Erize/B. Coleman Inc.
53(B) Francisco Erize/Bruce Coleman Inc.
54 Schraml/Jacana
55(TR) Francisco Erize/B. Coleman Inc.
55(BR) Christiansen/Frank W. Lane
56(L) Karl Maslowski/Photo Researchers Inc.
56(BR) Visage/Jacana
57 C. J. Ott/Bruce Coleman Ltd.
59 Philippa Scott/Photo Researchers Inc.
60 Robert Estall
61(R) W. Lummer/ZEFA
62 G. Maxwell/British Antarctic Survey
63(T) W. N. Bonner/British Antarctic Survey
63(B) M. R. Payne/Bruce Coleman Ltd.
64 Bill Wilson/Photo Researchers Inc.
65(T) S. Roberts/Ardea, London
66 C. J. Ott/Bruce Coleman Ltd.
67 Ardea, London
68(L) Frédéric/Jacana
68(R) Strobino/Jacana
69(T) N. Lightfoot/Photo Researchers Inc.
69(BL) Hugh Maynard/B. Coleman Ltd.
69(BR) Stephen Collins/Photo Researchers Inc.
70(T) Lelo/Jacana
70(B) Bill Wilson/Photo Researchers Inc.
71(BL) Mason Rutherford/Photo Researchers Inc.
71(BR) S. Roberts/Ardea, London
72(T) Varin/Jacana
72(B) Chris Knights/Ardea, London
74 Kenneth W. Fink/Ardea, London
75(TR) Bruce Coleman Ltd.
75(B), 76 Philippa Scott/Photo Researchers Inc.
77(T) Gordon Langsbury/Bruce Coleman Ltd.
77(B) J. B. and S. Bottomley/Ardea, London
78(T) W. Lummer/ZEFA
78(B) W. Ferchland/Bruce Coleman Inc.
79 Jonasson/Frank W. Lane
80 Francisco Erize/B. Coleman Inc.
81(T) D. & K. Urry/Bruce Coleman Ltd.
81(BR) Dr. P. Germain/Ardea, London
82 W. Curtsinger/Photo Researchers Inc.
83(L) Francisco Erize/B. Coleman Inc.
83(R) M. G. Richardson/British Antarctic Survey
84–5 Andrew Clarke/British Antarctic Survey
86–7 Clem Haagner/Ardea, London
88 Guy Mannering/B. Coleman Inc.
89(T) Francisco Erize/B. Coleman Ltd.
89(B) Wm. R. Curtsinger/Photo Researchers Inc.
90–1 Philippa Scott/N.H.P.A.
93 P. W. Bading/ZEFA
96(TL) Inigo Everson/B. Coleman Ltd.
96(TR) Carré/Jacana
96(B) G. Maxwell/British Antarctic Survey
98 W. Ferchland/ZEFA
99(R) Inigo Everson/British Antarctic Survey
100 Ian Collinge/British Antarctic Survey
101(T) D. G. Bone/British Antarctic Survey
101(BR) E. Mickleburgh/Ardea, London
103 P. W. Bading/ZEFA
104 Carleton Ray/Photo Researchers Inc.
105 Fred Bruemmer
106(T) Frédéric/Jacana
106(B) Spectrum Colour Library
107(T) M. G. Richardson/British Antarctic Survey
107(B) Ian Collinge/British Antarctic Survey
108(T) Clem Haagner/Ardea, London
108(B) Carleton Ray/Photo Researcher Inc.
109(T) Clem Haagner/Ardea, London
109(B) William Curtsinger/Photo Researchers Inc.
110(T) Kellett/British Antarctic Surve
110(B) M. G. Richardson/British Antarctic Survey
111(R) William Curtsinger/Photo Researchers Inc.
113 Al Giddings/Bruce Coleman Ltd
114–6(L) Institute of Oceanographi Sciences
116(R) Russ Kinne/Photo Researcher Inc.
118(L) G. Williamson/B. Coleman Ltd.
119 Dan Guravich/Photo Researcher Inc.
120 G. Williamson/B. Coleman Inc.
122 Fred Bruemmer
124(L) Hed Wiesner/ZEFA
125 J. & S. Brownlie/B. Coleman Ltc
127 *Sunday Times Magazine*
128 Ministère de la Guerre, Paris Photo Giraudon
130 The Mitchell Library, Sydney NSW/Photo Brian Bird
133(T) Brown Brothers
133(B) The Beinecke Rare Book and Manuscript Library, Yale, courtesy of American Heritage
134 The Parker Gallery, London
135(T) Aldus Archives
135(B) Photo Bryan Alexander
137 Peter Lake/Sea Library/Tom Stack & Associates
139 Steve Northup, *Life* © Time Inc

Artist Credits

© Aldus Books: Alan Hollingbery 10–1
David Nockels 50–1, 94–5

FRONTIERS OF LIFE

Part 2
Mountain Life

by Bernard Stonehouse

Series Coordinator	Geoffrey Rogers
Series Art Director	Frank Fry
Design Consultant	Guenther Radtke
Editorial Consultant	David Lambert
Series Consultant	Malcolm Ross-Macdonald
Art Editor	Susan Cook
Editor	Damian Grint
Research	Barbara Fraser

Contents: Part 2

Introduction

"Mountains" said Ruskin, "are the beginning and the end of all natural scenery." Most striking of the world's features, mountains dominate the skyline wherever they appear, raising our eyes above the everyday levels of civilization, dwarfing man, and trimming his grandest achievements to size. Everyone responds to mountains: to some they are hostile or awesome; to others challenging, protective, inspiring, to no one dull.

This book is about mountains, not as seen by poets or other men, but as living communities of plants and animals. Rising through the lower levels of the atmosphere, mountains provide a wide range of climates, drainage patterns, soil structures, and other conditions. These, in their turn attract a variety of life forms. Yet from the countless spores, seeds, and creatures that stray onto the slopes, the often harsh mountain environment selects rigorously. Mountain species—especially on steep, high peaks—are largely those that have survived severe natural tests, becoming adapted to highland ways in the process.

They are interesting and sometimes curious organisms, these mountain dwellers, drawn from diverse parts of the plant and animal kingdom and showing varying degrees of adaptation and specialization. Some—like the ibex, lammergeier, pika, and stone pine—are true mountain forms, well designed for life among the crags and rarely seen elsewhere. Others—like the porcupine, beaver, viper, panda, and mallard—are versatile species whose needs may be met on high ground or low. Some organisms—like the saxifrage of the Alps and the vicuña of the Andes—are clearly at home in an upland setting. Others—like the black-backed gulls, oceanic petrels, and parrots of New Zealand mountains—seem out of place on the heights. In these pages we see how such dissimilar plants and animals go to build the communities that make up the beautiful and exciting world of mountain life.

The Mountain Environment

By dictionary definition a mountain is an elevated portion of the earth's crust that rises conspicuously above its surroundings. Whether less than 1000 feet or more than 15,000 feet high, whether springing directly from sea level or from a high plateau or sierra, high ground matches the definition so long as it stands free above the surrounding land, and its head forms an identifiable landmark. Like most technical definitions, this is true but minimal. If mountains had no other qualities beyond elevation and conspicuousness, only climbers and surveyors would find them exciting. It is a dull, pedestrian description of the world's most splendid natural features, which have struck wonder into the mind of man and captured his imagination since he first raised his eyes to consider them.

Man seldom much exceeds six feet in height, and practically every elevation of the earth's surface towers above him. Yet early cartographers gave the names "mountain" and "mount" to high peaks and low, with a curious lack of discrimination. Mariners, accustomed to the broad plains of the ocean, tended to call any small coastal feature a mount. Mont-Saint-Michel, about 260 feet high off the Manche coast of France, would hardly be called a mountain in any other setting. The gentle, rolling Sussex Downs of Southern England were a "chain of majestic mountains" to the 18th-century naturalist Gilbert White, who seldom strayed far from his Hampshire parish and probably had little to compare them with. In similar vein, George Washington's Virginian hilltop home was imposingly named "Mount Vernon."

Today we expect more of mountains. Nearly everyone in Western society has seen the Alps or Pyrenees, the Rocky Mountains, Andes, or Himalaya, in reality or pictures, and we all have a much clearer idea of what true mountains should look like. To be worthy of the name, mountains should stand at least 1500 to 3000 feet above the surrounding country, forming a

Mount Eisenhower's naked rock walls soar over 9000 feet, dominating gentler, forested slopes rising from Bow River. The ingredients in this scene from Alberta's Banff National Park form part of the popular image evoked by the word "mountain."

Lowlanders miscalled many a trivial hump a mountain. This is Saint Michael's Mount, a rock lying off Britain's Cornish coast.

distinctive block of high ground with foothills about them. We tend to visualize such blocks as ranges dissected into matching peaks, though we recognize isolated volcanic cones as splendid exceptions to this general notion. Remoteness and ruggedness are part of the popular image, and indeed mountains of character have a quality of inaccessibility that sets them apart from man and his ways. Ideally they arise from wilderness and are skirted by dark, forbidding forest, which contributes to their isolation. We picture mountains as typically soaring up from the forest to well above the timberline, with zones of alpine scrub and tundra (regions of permanently frozen subsoil) and a capping of ice. Finally, we think of true mountains as having a fauna of their own—a community of distinctive animals selected by nature to withstand the hardships that, rightly or wrongly, we associate with mountain living. Mountains with some or all of these qualities are to be found on every continent. Majestic, mysterious, and often remarkably beautiful, they form the background to this book. Their communities of plants and animals, similar in outline but differing in detail on each of the continents, fill out the foreground.

Vast though they appear from the surface of the earth, mountains are in fact almost negligible bumps on the face of a smooth planet. Viewed from space, they disappear completely against the vast sweep of the earth's curvature. There are some 25 major blocks of mountains dotted about the continents—more if we include the half-dozen or so buried ranges of the ice-ladened Antarctic continent—and vast ranges of submerged mountains cross the floor of the major oceans. Yet the mean height of the continents above sea level is only about 2400 feet and less than one twentieth of their total area rises above 6000 feet. From the highest peak of the Himalaya (29,028 feet above sea level) to the greatest depth of the ocean (36,198 feet below sea level) is a mere 12 miles—less than one third of one per cent of the earth's radius. So the high mountains that dominate the skyline are, on a world scale, no more than trivial, cloud-catching irregularities. On a man-sized model of the globe, about six feet in diameter, the oceans could be represented by a film of water on average one twentieth of an inch deep. Mount Everest and its companions—the vast block that some call the roof of the world—would stand no more than the thickness of a slim coin above sea level. The Alps,

Andes and Rocky Mountains, with highest peaks between 15,000 and 23,000 feet, could be shown at their correct scale by cut-out shapes of fairly thin cardboard, with a sprinkling of coarse sugar to represent the peaks. The lesser mountains of Britain and mainland Europe would lie within the thickness of stout paper, and it would be difficult to make the model smooth enough to represent faithfully the small hills and valleys of the lowland surfaces of the continents.

The traveler from outer space, seeing mountains as trivial bumps on the earth's surface, might be excused for thinking that they could exert little effect on the plants and animals that live on them. We who are earthbound know better. Differing physically from their surroundings in structure, soils, aspect, climate, drainage, and many other important environmental qualities, mountains provide a range of challenges and opportunities for organisms—a range, in fact, of ecological niches—quite distinct from those of the plains. Often their slopes are more difficult to colonize, their soils thinner and drier, their climates harsher than those of low-lying land. But plants and animals that can cope with these conditions find advantages of space, relative freedom from predators, and the lack of competition denied to them in the crowded, cutthroat conditions of the plains. Attracting some organisms and defeating others, selecting rigorously among hundreds or thousands of contending species, mountains have gradually acquired and molded their own distinctive communities of flora and fauna. Although basically similar the world over, these communities of mountain life differ subtly from continent to continent, from range to range, indeed often from peak to peak within a single range.

What is special about mountain environments? How do they differ from situations on the plains close at hand, and how are mountain plants and animals adapted to them? To answer these questions we have to think first of the physical and biological factors that make up environments, directly and indirectly affecting the daily lives of organisms and determining the quality of their living. Atmospheric pressure and temperature, radiation, cloud cover, humidity, rainfall, snow, and wind are some of the atmospheric factors that change with increasing altitude and must be taken into account. More literally down-to-earth qualities include aspect in relation to the sun; angle of slope; soil porosity, stability, and fertility; drainage; vegetation cover and annual persistence of snow cover. All these directly affect the selection and survival of organisms that try to establish themselves in mountain environments. Then we must consider the repertoire of responses that organisms display in response to environmental challenges. Ability to acclimatize in the short term, and to adapt and remain adaptable in the long term, are perhaps most *general* characteristics of mountain plants and animals, but each species has its own *particular* combination of qualities that gives it success in its own environment.

First, let us consider the part played by the atmosphere. The atmosphere is a thin, nebulous veil of gases, pale blue in color, that covers the earth and fits tightly into every nook and cranny of its surface. The bulk of the atmosphere—roughly four fifths of it—is made up of nitrogen, a relatively inert gas: it combines reluctantly with other chemical elements. Although every living organism is packed with chemical compounds of nitrogen, few absorb the gas directly from the atmosphere. Rather more than one fifth of the air is oxygen, a lively and reactive gas that most plants and animal use daily throughout their lives for the vital process of respiration. A small balance of the atmosphere includes relatively tiny amounts of water vapor, ozone, carbon dioxide, more than a dozen other gases, and a wide selection of tiny particles that help to give the atmospheric veil its distinctive bluish haze. Water vapor and carbon dioxide circulate constantly between living organisms and the atmosphere; ozone, present mostly in minute quantities in the upper levels plays no direct part in the lives of plants and animals, but helps to protect them from the high-energy radiation beamed in from space.

There is no firm outer boundary where atmosphere stops and outer space begins. But over half the mass of the atmosphere lies below 18,000 feet (almost the height of Africa's highest peak, Mount Kilimanjaro). Over three quarters of the atmosphere is below 29,000 feet (about the level of Mount Everest). Practically all of it lies within 100 miles of the earth's surface, forming a layer just under four fifths of an inch deep on our man-sized model of the globe.

The atmosphere is densest at sea level, where the earth's gravity exerts its strongest pull, and very much rarefied at higher latitudes. Its pressure, measured by the height of the column of mercury it can support in a barometer tube, varies accordingly. At sea level, on average, atmospheric pressure supports a column 30 inches high. On the top of Kilimanjaro it can support only 15 inches of mercury, and on Everest less than 10 inches. As atmospheric pressure falls with altitude, so do the partial pressures exerted by the individual gases that make up the air, though the relative proportions of most of them stay remarkably constant. At the peak of Mont Blanc, nearly 15,800 feet high in the European Alps, each lungful of air taken by a climber contains only about half the oxygen that, volume for volume, he would obtain at sea level. On Everest only about one third as much oxygen is available—far too little to support an active man for any length of time. So at altitude there is less oxygen available for the respiration of plants and animals, less carbon dioxide available to plants for the food-making process of photosynthesis, and often—especially at levels above the clouds—less water vapor to provide the damp atmosphere that many plants and invertebrate animals require for their comfort.

Although atmospheric pressure and the partial pressure of vital gases change strikingly with altitude, surprisingly few kinds of organisms seem to be affected directly. There is little evidence that plants are affected at all, presumably because most of them make oxygen during the day, and during the night have a low rate of *metabolic activity* (the rate at which their life processes operate). Reptiles and amphibians, like invertebrate animals, also tend to live at modest rates, especially where air and ground temperatures are low. They are unlikely to run short of oxygen even at great altitude. But few reptiles or amphibians live at great heights in the mountains; they are too rapidly immobilized by cold, and would be unlikely to survive the long winters of the temperate and polar mountain ranges. However, insects adapt well to life at high altitudes, and many kinds are mountaineers. Other invertebrates are less successful, but more likely to be affected by cold and aridity than by low atmospheric pressure and shortage of oxygen.

Birds spend much of their lives looking down upon the rest of the world, and seem well able to cope with low atmospheric pressures. They are active at all levels of the mountains in summer, when living conditions are at their best and food is plentiful. Many species of small birds breed successfully far above the timberline

The special living conditions that mountains afford and deny have weeded out all but those plants and animals able to cope with the rigors that go with high-altitude life. These durable organisms range from big, sure-footed mammals such as the Rocky Mountain goat (above), to such tiny plants as Paraquilegia grandiflora, *here photographed blooming 14,000 feet up among Himalayan rocks.*

on every continent, at heights where few mammals of any size are found. Tawny eagles, lammergeiers, and other large, soaring predatory birds scan the high plateau of Ethiopia from heights above 13,000 feet, and condors have been reported above 23,000 feet over the Andes. Choughs breed above 16,000 feet in the Himalaya, and people have seen them at over 26,000 feet on the slopes of Mount Everest. Birds live energetic lives, with high and constant demands for oxygen to maintain their high rate of metabolism. They possess a rapid heartbeat, remarkably efficient breathing apparatus and rich hemoglobin (an oxygen-absorbing pigment in the blood). Birds may well owe success at high altitudes to this efficient combination, which at low atmospheric pressures would help them to trap and hold whatever oxygen was available.

Mammals, too, have a high metabolic rate; but while their lungs and blood system take them comfortably to elevations of 6000 to 10,000 feet, few seem disposed to live at higher altitudes, and those that do show interesting degrees of physiological adaptations for coping with the lack of oxygen. Some kinds of herbivores make annual migrations from lowlands to highlands, spending their winters grazing in the warm valleys and their summers on mountain pastures

The higher you climb, the thinner the air, and the more difficult breathing becomes. Tenzing Norgay and Edmund Hillary (above) had to wear oxygen masks to conquer Mount Everest, more than 29,000 feet high. No such aids are needed upon the Swiss-Italian Matterhorn (right), barely half the height of Everest.

thousands of feet higher. For them a period of acclimatization—physiological adjustment over a period of several days or more—is usually necessary. This ensures that a sudden emergency (perhaps the need to fight or escape from a predator) does not find them breathless and unable to exert themselves to full capacity.

Different mammals acclimatize to different degrees, using different combinations of physiological mechanisms. As a general rule their bodies respond to the move up into the mountains by generating extra blood cells and more hemoglobin, causing a slight increase in total blood volume. The heart rate gradually speeds up, blood pressure rises slightly, and the normal rate of breathing may increase. But species vary in their adaptability. Domestic cats, for instance, seem poorly equipped for altitudinal changes. They reportedly become miserable when carried to heights of 10,000 to 13,000 feet, sometimes suffering convulsions and even dying. Dogs, mice, and men in similar circumstances adjust readily, though sometimes passing through a short period of "mountain sickness." In man this involves spells of heartpounding and breathlessness, loss of appetite and of sleep, also coughing, nausea, headaches, muscle cramps, and a general feeling of malaise and irritability. Properly acclimatized, most people can live comfortably for long spells at heights of 8000 feet or more, and many lowlanders have spent several weeks at a time at 16,000 to 20,000 feet on climbing and research expeditions. But above 20,000 feet man cannot live for long without physical deterioration.

Mammals that always live high up tend to have more red corpuscles and slightly larger hearts than their nearest lowland kin, both presumably in compensation for the lower oxygen concentrations in which they are living. Some, such as the South American vicuna, possess unusual forms of hemoglobin that show a stronger-than-normal affinity for oxygen; and llamas, tahr (Himalayan mountain goats), and Barbary sheep circulate individual red corpuscles longer than lowland mammals. Men who live permanently at high levels find no difficulty in working strenuously there. Compared with lowlanders they breathe faster, have more oxygen-carrying pigment in their blood and, like the vicuna, may possess blood pigment with a slightly stronger affinity for oxygen. Perhaps most important of all, they have a philosophy of life that allows them to stop working when they are tired, a fact that acclimatizing lowlanders often find puzzling. Whether some or all of the mountain-dweller's physiological characteristics are inborn or acquired during life is not entirely clear. Newborn babies seem to lack them, and they are similar to some of the qualities acquired during acclimatization. So there are strong suspicions that, physiologically speaking, mountain folk are basically similar to lowlanders.

So far we have seen how atmospheric pressure influences mountain life. But pressure is only one of the atmospheric factors that changes with increasing altitude. Temperature is another, perhaps more directly involved in the lives of most mountain plants and animals. Temperature depends ultimately on radiant energy from the sun. Because the atmosphere is transparent to most of the energy arriving from the sun, much of this energy reaches the earth's surface, where part is converted to heat. Radiation, convection and conduction transfer much of this heat to the

Air gets both thinner and colder as altitude increases. Thin air makes a poor trap for the sun's heat. Thus even near the equator, snow crowns the tallest peaks. Above: beyond these giant groundsel and lobelia plants, snowfields cling to the summit of Mount Kenya, an extinct East African volcano almost astride the equator. Right: in response to the thin air of the high Andes these Andean Indians have developed roomy lungs and their blood makes more use of available oxygen than lowlanders' blood does.

atmosphere, which traps some of the heat, thus acting as a greenhouse in tropical and temperate latitudes and as a warming blanket in colder parts of the world. Where mountains rise high into the upper regions of the atmosphere, the greenhouse is less effective and the blanket thinner. Mountain surfaces thus grow consistently colder with altitude. As a rough guide, mean air temperature falls 3.3°F for every 1000 feet increase in height above sea level. It is thus not surprising to find permanent snowfields and solid ice capping many tropical mountains—Mount Kenya, for instance, which stands astride the equator in East Africa and is ice-bound above 15,000 feet.

The thin mountain atmosphere affects environmental temperatures in other ways too. Because radiant energy from the sun has relatively little atmosphere to traverse, relatively little of the energy is absorbed on the way. So the full force of the sun is felt at high altitudes, particularly on sun-facing slopes, where the ground surface may get very hot during the daylight hours. At night the thin atmosphere makes a poor blanket for the mountain slopes, and the ground heat is radiated rapidly back to space. Mountain habitats therefore tend to suffer extreme daily fluctuations of temperature. Close to the snow line freezing and thawing may each occur once every 24 hours for most of the year—an uncomfortable business that few organisms can tolerate. Even during daylight hours differences between air temperature and ground temperature may be extreme. Above about 13,000 feet on Mount Kilimanjaro on a sunny day the ground may be as much as 50°F warmer than the air three feet above it. But ground temperature falls rapidly when the sun is hidden by cloud.

Plants living high on the mountains tend to grow short and straggling, their branches hugging the ground to keep as close as possible to its warmth. Like the animals that live among them, they must be capable of withstanding both a wide range of temperatures and sudden changes. The most successful alpine plants include many sedges and grasses, saxifrages, gentians, pinks, and small compact shrubs of the rose family, some familiar to rock gardeners all over the world. These are members of plant families well represented in both alpine and polar regions, where extreme cold occurs for at least part of each year. Alpine animals respond in different ways to the low temperatures that strike them as soon as the sun goes down. When the morning sun disappears behind a cloud, lizards that were warming themselves sneak quickly back to their crevices to avoid the sharp though temporary chill that follows. Insects tend to be small and they are active only during the warmest part of the day—possibly for a few hours in summer, a few minutes in spring and autumn. Birds and mammals, too, tend to be diurnal, relying on their high metabolic rates to generate heat and on their thick plumage and fur to retain it. To cut down radiation heat loss at night, small mammals burrow or hide among rock crevices; larger mammals seek shelter under overhanging rocks or trees.

Seasonal temperature changes, caused by the sun's apparent annual migration from tropic to tropic, are barely felt on the tops of equatorial mountains. As several researchers have pointed out, temperature differences between day and night are far more marked than differences between summer and winter; on the upper slopes of Mount Kenya it is "summer" every day and "winter" every night in July, November, and March alike. But in temperate and polar regions the effects of seasons are strongly felt, and indeed exaggerated, on the mountaintops. Because it is colder throughout the year, the season of growth and reproduction for plants and animals of the highlands is shorter than down in the lowlands, and the most successful organisms have adapted to this harsher regime. Most flowering plants of the mountains are perennials, which store reserves of energy within the plant body over several seasons before flowering. Most of their reserves are held in extensive root systems and

underground stems, bulbs, and corms, where they are protected from frost and predators. Many plants reproduce vegetatively rather than by seed, putting out runners to form clumps and colonies. Some, such as the drooping saxifrage, form axillary bulbs—tiny, self-contained buds that drop to the ground close to the parent plant and sprout independently. Others, such as the mountain poa grass, produce seeds that germinate on the stem, and drop to the ground as small readymade plants. Mountain animals are on the whole less ingenious in their adaptations. Their response to the short mountain season is to produce few young by restricting the size of their litters, the number of litters produced, or both. This is especially noticeable in mountain rodents: whereas their kin on the plains raise several large litters each year, those adapted to mountain life seldom produce more than one small litter each season.

Survival in prolonged periods of cold poses a relatively straightforward problem for mountain plants and animals. But the combination of cold with high winds is especially hostile to life, and mountains are often battered by blustering turbulent winds blowing from every corner of the sky. In lowland areas and rolling, undulating country, moving air is slowed significantly by friction as it passes over the ground. But among high peaks there is less to hold it back, so the upper slopes of mountains tend to be windy. Systemic winds—winds due to the movements of cyclones, anticyclones, and other shifting atmospheric pressure systems—reach their greatest strengths at high levels; a steady breeze across the plains may be the only indication of half a gale or more in the highlands above. To add to these hazards and discomforts, mountains generate winds of their own. *Katabatic* winds are masses of cold air rolling downslope under their own weight. In gentle form they occur almost any evening in any open valley, as the sun departs from the flanks of the hills and cool air gathers on the valley floor. In wilder form they shriek down the valleys and across the foothills, sometimes warmed by compression and accelerating as they go. These large-scale katabatic winds usually occur on the leeward side of a mountain range when a difference in pressure across the mountains topples the local stability of the atmosphere. The *föhn* winds of the Alps, the *chinook* of the Rockies, and the *nor'westers* of New Zealand's Canterbury Plains are katabatic winds of this kind. Because they have lost most of their moisture on the windward side of mountains they tend to be dry; often they are hot winds, blowing persistently for days on end. Warm-blooded animals, including cattle, sheep, and people, find them intensely trying. The incidence of attempted murders, marital bickering, and automobile accidents in small mountain communities is said to rocket upward when the katabatic winds are blowing.

Winds tend to exert a wide range of harmful effects on alpine plants and animals, though most communities living in windswept mountain regions suffer little; the process of natural selection ensures that they have already taken their precautions and developed their strategies of defense. Plants hug the ground tightly, or grow in compact stands that strengthen and

Survival on high mountains may hinge on ability to trap water and heat. Left: this plant 12,000 feet up on Mount Kinabula in Borneo catches rainwater channeled along a crack in the rock. Right: thick fur and its underground burrow conserve body heat for the collared lemming, a small North American mammal.

protect each other from the blast. Often on the most exposed patches of open ground they appear in characteristic "cushion" form—large colonies of tightly packed plants, all developed by budding from a single individual, offering a large surface of tiny green leaves to the sunlight and exposing a compact, united front to the wind. Even gentle winds affect plants significantly by tending to increase their rate of transpiration (loss of water) through the leaf pores or *stomata*. Plants respond by closing the stomata, but this restricts their capacity for photosynthesis and for transporting materials between stems, roots, and leaves. In summer the drying effects of light winds are seldom serious, because plants can usually draw water in through the roots to make up for any lost through the leaves. In winter, when the ground is frozen and water immobilized, even light breezes can kill leaves by drying them out completely. Violent winds are far more deeply destructive, for these tear up turf and shift soils, sometimes exposing the bare ground beneath to erosion during summer rains. All too often this happens in high-country pastures where man allows overlarge flocks of sheep, goats, and cattle to graze and browse; the first damage is done by too many sharp hooves and nibbling teeth, which scar the turf, open it, and let the wind in.

Animals react to mountain winds in various ways. Some tiny creatures use the winds as an agent for dispersal: spiders, caterpillars, and several other kinds of invertebrate float effortlessly from one mountain peak to the next on the turbulent air. Mosquitoes and flies tend to avoid strong winds, lying low among the vegetation while gales blow, and emerging to dance, buzz and bite during calm spells. Mountain deer and goats often avoid the attention of biting insects by seeking out windy headlands, where they may browse and ruminate in comparative peace. On hot sunny days exposure to mountain winds may be their only means of keeping cool. Mammals that winter in the mountains avoid the wind where they can. The smallest mammals, including shrews, mice, voles, and ground squirrels, spend the coldest part of the year under the snow, living on the previous summer's seeds and leaves.

Larger mammals—guanacos and vicunas of the Andes, yaks and snow leopards of the Himalaya, and the widely distributed mountain goats and sheep—grow thick winter coats, and often dig trenches or dens in the snow to avoid the main blast of a cold wind. Even the strongest winds find difficulty in penetrating the dense, matted wool of a yak, the lightweight but remarkably efficient fleece of a vicuna, or the dense, sleek winter fur of a mountain-dwelling leopard. But many grazing mammals, including caribou and deer, leave the windswept mountains in winter, retreating to the forests and valleys where the air is calmer and food more easily found.

Cloud cover, humidity, and radiation levels are closely linked factors of the mountain environment. Reaching high into the lower atmosphere, mountains rise through air that gains steadily in relative humidity as it grows colder with increasing altitude. At so-called cloud level the air is saturated, water vapor condenses into visible droplets, and heavy mist, rain, or snow swirl persistently. Turbulence may carry this band of saturated air high above the cloud base, to form a wet zone thousands of feet thick.

Cloud often forms a permanent or semi-permanent layer on the windward slopes of a mountain range, where the prevailing breezes constantly bring in more moisture for condensation. On many mountains in warmer tropical and subtropical regions of the world the broad band of saturated air gives rise to a distinctive zone of damp "cloud forest." Trees, shrubs, ground vegetation, and rocks are permanently wet, usually enshrouded in a mobile, writhing mist, that tropical man finds uncomfortably cold. Every surface tends to be covered with wet mosses, lichens, algae, and other *epiphytes* (plants that get nutrients and moisture from air and rain, and often grow upon other plants). Growth is often lush, though individual trees and shrubs may be locally stunted through lack of sunlight. In the eastern highlands of Africa annual precipitation increases up to about 4500 feet above sea level, that is, a few hundred yards above the base of the lowest clouds. The damp air supports a broad zone of wet montane forest at this level, and up to a height of 8000 to 10,000 feet. In the drier air beyond, forest gives way to heath and moorland, and eventually to grassland and alpine tundra. This is typical of many tropical regions in the world. In the European Alps, and probably in many other temperate and cooler subtropical regions, the zone of heaviest annual precipitation lies at about 11,500 feet above sea level. Here the air is cold for much of the year, and a high proportion of the annual precipitation falls as snow, remaining frozen for many months. So at these levels the climate is relatively dry, despite the precipitation figures. The combination of persistent cold and relative dryness make it impossible for forest to grow, and the characteristic vegetation tends to be heath and alpine meadow, rather than cloud forest.

Above the cloud level in tropical or temperate regions the sun shines brilliantly through a thinner, drier, and cleaner atmosphere. Because it lacks water vapor, dust, and pollen, high mountain air tends to be crystal clear, allowing the splendid visibility we usually associate with mountain regions. The air here also lets through heavier concentrations of solar radiation than the denser atmosphere at lower levels. In the European Alps, places at about 10,000 feet receive one fifth more sunshine in summer, and one third more in winter, than settlements in the valleys close to sea level. High intensities of solar radiation, coupled with low night temperatures, tend to slow down the rate at which plants grow. This is why many mountain plants have the short stems and squat, stunted form that serve them so well in exposed situations. Also, although sugars created in leaves and stem during the intense photosynthetic activity of the day cannot be transported away at night, their accumulation may help to protect the cells from freezing. It may also produce the unusually vivid pigments that color the flowers and leaves of alpine plants in the late summer and fall.

Solar radiation received at high level may contain as much as 50 per cent more in the ultraviolet range than radiation received at sea level. This radiation, partly direct from the sun and partly reflected by snow surfaces, gives alpine skiers their healthy tan. But it may help to cause the high incidence of skin cancers suffered by mountain folk who frequently expose their skins to the unshaded rays of the sun. However, it is not at all clear whether the very high incidence of ultraviolet and other radiations has either a harmful or beneficial effect on plants and animals of the mountains. Some biologists think that the truly harmful amounts of radiation are all filtered out by ozone in the upper atmosphere, far above the highest peaks. If this is so, plants

Many mountain plants and animals have become adapted to survive specific hazards. Like some other black creatures of the high mountains, this alpine chough (above) may find its color offers some protection against damage from intense incoming radiation. The low profile of moss campion (below, photographed on Ben Lawers in Scotland) helps this plant to escape the worse effects of fierce mountain gales. Certain plants, even more exposed cheat the wind by assuming the form of ground-hugging cushions.

High peaks are colder than lowlands yet, paradoxically, may receive the sun's rays in larger doses. High among Italy's Dolomites, skiers warmly wrapped against cold sit in snow to get suntanned faces (left). But intense radiation or cold can make summits biologically sterile, as in the Alaska Range (right).

and animals are safe from radiation hazards, no matter how high they live. But other biologists feel that the relatively dark pigmentation of many mountain butterflies and birds, and the tendency to have more chromosomes than their lowland counterparts are protective responses to high radiation levels. This would suggest that the effects of radiation are far from negligible. Experts have also suggested that the total lifelessness of very high mountain slopes, and the thin vegetation in the uppermost regions of the alpine tundra, are due as much to the damaging intensity of this radiation as to the intense cold, strong winds, and aridity. Few biologists have lived and worked on the mountain tops long enough to have investigated this interesting problem in any depth.

Soils provide the medium in which mountain vegetation grows. In many ways they determine the wealth of vegetation cover, and so control the wealth and density of animal life. Let us therefore now look at the soils and their effect on life in the mountains. By comparison with the lowlands, mountain soils are usually poor. With few exceptions they tend to be young and immature, only recently derived by weathering from the parent rock and not yet yielding up their minerals to plants in soluble, readily available form. Often the weathering process is continuing so that new, insoluble particles form a high proportion of the soils. Usually they are subject to constant sorting and re-sorting by winds, rain and frost-heaving, so that they seldom have time or opportunity to mature. Most mountain soils lack humus—the rich organic component of fertile soil. This is because plant growth on mountains is generally sparse, especially at the highest and coldest levels, and little organic material collects to be broken down into humus by soil organisms. Most soil organisms are anyway ill-equipped for the high-level mountain environment, with its scorching days, cold nights, and lack of protective ground cover.

Water, or lack of it, powerfully affects the nature of mountain soils. Although the overall slope of a mountain surface encourages the rapid run-off of rainwater, not all mountain slopes are well drained. Alternating layers of permeable and impermeable rocks often create pockets and broad expanses of marshy ground. Springs, streams, tarns, bogs, and sodden turf are common features of mountain slopes in the damper regions of the world, together with wet, acid soils that relatively few species of plants tolerate. On the other hand, where there are uniformly permeable rocks such as limestone, chalk, and sandstones, rainfall and surface

moisture soak very quickly into the ground. The thin soils dry out rapidly between showers, and vegetation is often restricted both by chronic aridity of the soil, and by mineral deficiencies resulting from the constant leaching—a process in which rainwater dissolves soil minerals and washes them downward. Because leaching proceeds faster than soil particles break down, mountain soils often lack essential minerals that abound in the rocks from which they were formed: as many hill farmers know, chalky soils may need frequent liming to keep them fertile. So a relatively small area of mountainside may hold patches of soil impoverished by leaching and aridity, only a few yards from sour, waterlogged marsh and wet heathland. The differences often show clearly in patterns of vegetation, which vary strikingly from patch to patch.

Some of the most convincing mountain soils occur below cushion plants, among the roots and

Above: hillside trees, pastures, and bare rock form a patchwork vegetation pattern in Washington's Olympic National Park. This mountainside plant variation over small areas can reflect local differences in soil, rock, and water content.

Snow can benefit plants high on the mountains. Above: in spring, melting snow provides moisture for flowering alpine plants like this Soldanella. *In winter, snow protects vegetation from frost and the drying effects of wind. Below: a Swiss scene shows that even needle-leaved conifers may support snow.*

matted lower branches of tiny shrubs and ferns, in tussock stools, and in crevices between rocks where ferns, grasses, sedges, and small flowering plants have built up a home for themselves. Rooted plants, unlike the simpler mosses, lichens and algae, provide mats of fibrous roots in which minerals, moisture, and soil organisms collect. Under the cover of the plants, these key elements gather in sufficient quantity to interact upon each other and form small pockets of brown, earthy soil. While nearby surfaces are pounded by rain, scorched by the sun, and heaved up by frequent frosts, the protected patches mature in peace, building up thriving populations of a wide variety of algae, fungi, bacteria, worms and other small creatures, whose daily business is to manage and process organic material. Maturing soil attracts more plants, so that, over the years, large areas of mountain slope may acquire stable tracts of turf, with deep-rooted shrubs and tall tussocks. However, even the most stable, mature and permanent-looking patches may be stripped by the winds of a single hard winter, or flushed downhill by a change in surface drainage patterns, allowing the whole cycle to start again.

By far the poorest mountain soils occur on the steepest slopes, especially on sunless slopes, which tend to be cooler and less protected by vegetation than sunny slopes. But the barrenness and inaccessibility of shady slopes is put to good use by the animals that live on them. Steep cliffs and bare rocks are a haven for mountain sheep and goats, whose sharp, cloven hooves are cleated like expensive climbing boots to grip wherever they can. Few predators care to follow these browsing and grazing animals into the fastnesses of the higher slopes, where there is no cover for stalking, the ground itself is treacherous and scent carries unimpeded on the wind. Smaller mammals—ground squirrels, pikas, and marmots, for example—benefit by burrowing into the rocky ground or finding safety in the tunnels and cavities of *scree*: loose, frost-shattered rock debris. These are situations where bears, foxes, jackals and other predators are at a disadvantage; they are too large to follow their prey in, and often not strong enough to dislodge the heavy rocks that often protect them. Many mountain birds nest underground, including South American shaketails and earthcreepers, European wallcreepers, and several species of oceanic petrel, which breed in the alpine zones of coastal mountains in New Zealand, Antarctica, and South America.

The best mountain soils are those of high alpine meadows and parklands, where the ground is level or gently sloping, moist but well drained, sunny and warm during the growing months of late spring and summer, and well protected by valley walls and headlands from the searing effects of wind and frost. These soils provide rich pasture for mountain animals of all sizes, from mice, voles, ground squirrels, and porcupines to deer, antelope, and sheep. They attract also a rich fauna of insects, which feed on the meadow plants and their products (pollen, nectar, and seeds), and flocks of birds, many of which migrate up from the lowlands each year to feed on the summer bounty of mountain lands.

Most mountainous regions experience heavy snowfall in autumn and winter. This produces extensive snowfields and drifts several yards deep that persist well into spring. At first glance snow would seem to be more of a curse than a blessing to mountain organisms: a cause of destruction and death over wide expanses of land. Certainly snow affects the life of any plant or animal that winters in high mountains anywhere in the world. Driven by fierce winds at temperatures well below freezing point, snow can be as abrasive and damaging as sharp sand. The cold blast of snow-filled winds in autumn may severely cripple and deform small trees at the edge of a mountain forest. Deep drifts keep many of the larger herbivorous animals from their food in winter, and unseasonal snow can kill mountain birds and mammals, or at least disrupt their breeding cycles in spring.

However, on balance, snow is a boon. Some mountain biologists would argue that, without annual snow cover, life above the timberline would be almost impossible in many parts of the world. Snow protects trees, shrubs, and herbs from mechanical damage by casting a soft but impenetrable blanket about them. It guards many small plants from the fatal drying effects of winter gales. In the same way, snow keeps the sharpest frosts of winter from living tissues, both above ground and in the soil. Small burrowing animals live actively under it, spending much of their winters at the interface between ground and snow, where they scavenge for seeds and dried-out vegetation from last year's harvest. Though large snow banks inhibit the growth of some of the plants they cover, in melting they meter out moisture over several weeks to plants growing around their edges, and prevent the too-rapid warming and drying-out of the ground in spring. Rooted plants growing at extreme elevations of 20,000 feet and more in the Himalaya generally crouch in snow-crannies—small, sheltered places where soil has accumulated, and where snow provides both protection in winter and a steady trickle of water in spring and summer. The harshest conditions of all occur on the bare tops of mountains where snowfall is meager, winter snows cannot collect, and the ground remains exposed to the worst extremes of climate throughout the year.

Vegetation cover itself has a softening influence on the mountain environment. Wherever the ground is protected by even a thin layer of vegetation, soil temperatures and humidity are stabilized, windspeeds at ground level are reduced, there is less disturbance by frost and rain, and more opportunities arise for new organisms of all kinds to move in and multiply. Often a small patch of vegetation provides microscopic plants and animals with a congenially mild, damp, and windless oasis in little

Well-grazed pasture land in the French Alps. High alpine meadows provide some of the richest of all mountain soils, a fact that mountain shepherds and herdsmen exploit.

danger of frost or other disruption. Scientists working 13,000 feet in the alpine zone of Mount Kenya, found that temperatures in the outer leaves of a grassy tussock ranged from 55°F in the heat of the afternoon almost to freezing point at night, but during the same period temperatures at the base of the tussock varied only between 48°F and 43°F. Small leaf-dwelling animals fed in the outer leaves only during the early mornings and evenings, and retired to the more stable environment of the base during the heat of the day and intense cold of the night. Most of the small species of mammals and birds, and many of the insects and larger invertebrates, live and feed only where vegetation provides continuous cover, food, and protection from the elements.

With so many hazards, it may seem strange that mountains attract so wide a variety of plants and animals. Yet only the newest mountains stand bare. Practically all mountains are clothed in a continuous and permanent mantle of vegetation, at least to the level of permanent snow. What are the origins of mountains and of their hardy colonizers? How do these organisms select and fit into their habitats? How are their communities organized, and how do these differ around the world? These are the topics of the rest of this book.

The Making of a Mountain

Inverted V rock formation in Morocco's High Atlas Mountains illustrates in miniature how the world's great fold mountains were raised. Viselike pressure from opposite sides have bent horizontal rock layers and thrust them upward. On a far larger scale, buckling built the Andes, Alps, Rockies, and Himalaya.

The world's two dozen or more major blocks of mountains are scattered about the continents seemingly at random. High chains of alpine mountains line the western flank of the Americas but not the east. Towering alps cross southern Europe but broad plains straddle much of the north. Tall, isolated volcanoes overlook Africa's eastern rift valley system—all of them relatively new, some still smoking—but ranges of folded mountains rise in the northwest. Curved chains of mountainous islands punctuate the western Pacific Ocean, but not its eastern shores. The Antarctic Ocean contains the curious mountain-island of South Georgia, little more than 10 miles long but rising from deep water to form a chain of alpine peaks—the tallest almost 10,000 feet high. The Himalayan region of Central Asia rises almost to 30,000 feet, with over 70 magnificent peaks above 21,000 feet. In South America the Andes attain nearly 23,000 feet with serried rows of snowcaps above 15,000 feet. In contrast, the whole vast continent of Australia has nothing much above 7000 feet, though New Guinea, its island neighbor to the north, is crowned with peaks over 15,000 feet high.

Similarly the distribution of mountain life seems arbitrary, with no two plant and animal communities quite the same. A well-briefed spaceman, landing in the dark on the slopes of one of the world's higher peaks, should be able by morning to identify his position from the wildlife around him. In East Africa he might land in a dense garden of gigantic lobelias and groundsel, and descend through a forest of clattering bamboos, meeting gorillas, lions, elephants, and leopards on the way. South America would show him high, bare desert steppe of grass tufts and scattered woody shrubs, grazed by llamas, alpacas, and vicunas. Lower down are giant cacti, and a damp woodland of miniature, moss-clad trees ("elfin forest"), which in turn gives way to the richer tropical forest of the foothills—home of noisy, long-tailed monkeys and a bewildering host of colorful birds and insects. The Rockies would be tougher. Starting from snow-clad rocky tundra grazed by bighorn sheep, our spaceman would descend to rough scrub and meadow with deer, black and brown bears, mountain lions, hares, and blue grouse. Below, he would find dense forests of pine, fir, and spruce, with a rich fauna of small ground-living mammals. Australia's gum trees and marsupials would be an instant give-away. New Zealand's high tussock grasslands and dark, silent forests of evergreen beech would be quite distinctive. But their introduced fauna of merino sheep, Himalayan thar, Canadian wapiti, European red deer, chamois, and hares, Japanese sika deer, Indian sambar, and Australian opossums might well confuse the spaceman at first glance. In the eastern Himalaya he would find high alpine grassland and dense forests of Asiatic cedars, pines, and firs. He would descend through thickets of rhododendrons, azaleas, and bamboo, discovering colorful pheasants on their home ground, also musk deer, goat antelopes, tigers and one or both species of panda.

The distribution of the world's mountains and that of their plants and animals are in fact closely linked. To understand both, we must first look at the structure and movements of the continents that underlie the mountains and trace back their geological history.

To early geographers and laymen alike the physical world formed a constant, unvarying backdrop to the activities of man. Christians believed that oceans and continents, streams and lakes, hills, valleys, and plains had all been formed on the second and third days of creation, an event dated about 4000 years before the birth of Christ and only a few days before the advent of man himself. Except for the Flood, the face of the earth had supposedly changed little since then, and seemed likely to remain unchanged until the Day of Judgment. Rivers flowed endlessly, mountains stood timeless; tiny, transient man envied their immortality.

The notion of a static world remained widespread until the early 19th century, when quarrymen and engineers began to dig deep into the surface of the earth for coal, ores, and building stones. Many a quarry and pit, many a railroad tunnel and cutting, bared an orderly arrangement of distinctive rock layers. Clearly, such

Three mountain landscapes from widely separated parts of the world illustrate the seemingly random distribution of different mountain plants and animals. Above left: the snow-clad rocky tundra and coniferous forests of the Canadian Rockies. Above: llamas on the windswept grasslands of South America's high Peruvian Andes. Left: a New Zealand hillside peppered with the tussocky grasses that cover much of the South Island. Close study of these regions reveals good evolutionary reasons for the differences between their plants and animals—reasons rooted in the origins of mountains, in past climatic changes, and, especially in the cases of Peru and New Zealand, in the effect that man has had on them.

layers had been laid down as sediments in long-vanished lakes, estuaries, and oceans—sediments that had grown hard and compact with time. The layers contained the fossil remains of dead plants and animals, some similar to—indeed ancestral to—living forms, others extinct and with no obvious living relatives. Strata and fossils together told of a restless earth where oceans and continents had changed places, rocks had risen in folds under immense pressures to form mountain chains, deserts had become forests, and over immensely long periods of time both animal and plant communities had changed, keeping pace with the changing landscape. This mid-Victorian model of an evolving world revolutionized the thinking of philosophers of all kinds. The simple truth that the earth and its organisms had evolved through tens or hundreds of millions of years, and indeed were still evolving, brought new vitality to the natural sciences. Geologists, hitherto busy cataloging and classifying rocks, began to probe the mountain-building processes that were happening about them. Biologists increasingly turned from their museum studies of the anatomy and classification of plants and animals to investigate adaptation—how organisms had become fitted, and indeed were still fitting themselves, for the ever-changing environments in which they lived.

Back in the 19th century, geologists who studied the Alps, the Rockies, and the world's older mountains were well aware that parts of the earth's surface had repeatedly been thrown into folds by colossal compressional forces. But the true nature of those forces has only recently been made clear. Geophysical exploration since the late 1950s has shown us that the solid surface of the globe is actually a mosaic of thin, mobile plates, shifted around by convectional currents in the mantle—the hot, semi-plastic layer beneath the earth's crust. We now know that movement between the plates produces not only lateral pressure, uplift, and folding, but almost all the earth's major surface features, including rift valleys, volcanoes, island arcs, and submarine mountains and trenches.

Five of the mobile plates carry the continental masses, and it is their movements that produce the remarkable effect called "continental drift." First postulated in the early 1900s, the idea that continents moved in relation to each other was developed by the German geophysicist Alfred Wegener to explain many geophysical phenomena including the shape of the continents, mountain building, earthquakes and—at least partly—the world distribution of different plants and animals. Wegener and his followers piled up an impressive body of circumstantial evidence to back his hypothesis, but it has taken some of the latest geophysical discoveries to convince most earth scientists that the continents have shifted, and are still shifting at measurable rates. Biogeographers are delighted, for they can largely explain many puzzling distributions of plants and animals if it is allowed that the continents have strayed during the past 100 million years.

The six major plates and several minor ones that now make up the earth's crust drift from two fifths of an inch to four or more inches per year, and the present arrangement of plates and continents has evolved over the last 250 million years. In Permian times (280–230 million years ago), all the present continental blocks formed one vast supercontinent that Alfred Wegener called *Pangaea*. During the Triassic period, about 200 million years ago, Pangaea began splitting into two: *Laurasia*, a northern supercontinent including most of Europe, Asia north of India, and North America, and Laurasia's southern counterpart *Gondwanaland*, which included South America, Africa, Arabia, India, Ceylon, Australia, and Antarctica. Laurasia and Gondwanaland in turn started splitting, separating along lines that roughly match the present day continental coastlines, but more precisely follow the 1650-foot submarine contour. In fact the edge of the continental shelf bounds the fragments that were once Laurasia and Gondwanaland. Seas filled the spaces created between these fragments as plastic crustal material welled up from the earth's interior, re-forming the ocean floor and creating the sub-oceanic crustal rims of the plates bearing the continents. This upwelling process has gradually swung the plates and their continental blocks into their present

Right: this slab of imaginary land contains the major kinds of mountains; the cross sections help to show how they arose. Fracturing of the earth's crust along a fault (a line of weakness) produced the parallel fault block ranges on the left. Crustal buckling pushed up the fold mountains in the center; erosion has removed part of the outer rock layer, to give parallel hogback ridges. Right of the fold mountains stands a dome mountain formed from molten granite welling up between horizontal strata. Erosion bared the granite, leaving flanking ridges—remains of the strata once capping it. The volcano (far right) comprises solid lava plugs that filled the vents of a cinder cone, long since eroded. Faulting, folding, and volcanic action were responsible for the complex mountain range in the distance.

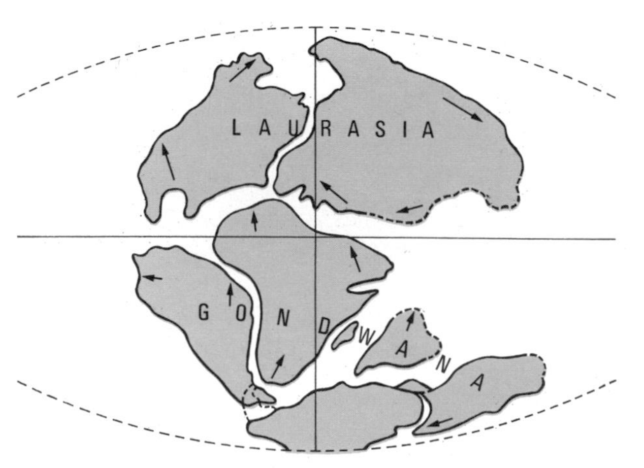

Left: reconstruction of the positions of the world's landmasses about 135 million years ago. The supercontinent Pangaea had split into Laurasia and Gondwanaland, divided by the Tethys Sea. Further splitting was to break Laurasia into North America and Eurasia, whereas Gondwanaland gave birth to South America, Africa, India, Antarctica, and Australia. Conflict between crustal plates carrying the continents was to throw up the major mountain systems on the earth today.

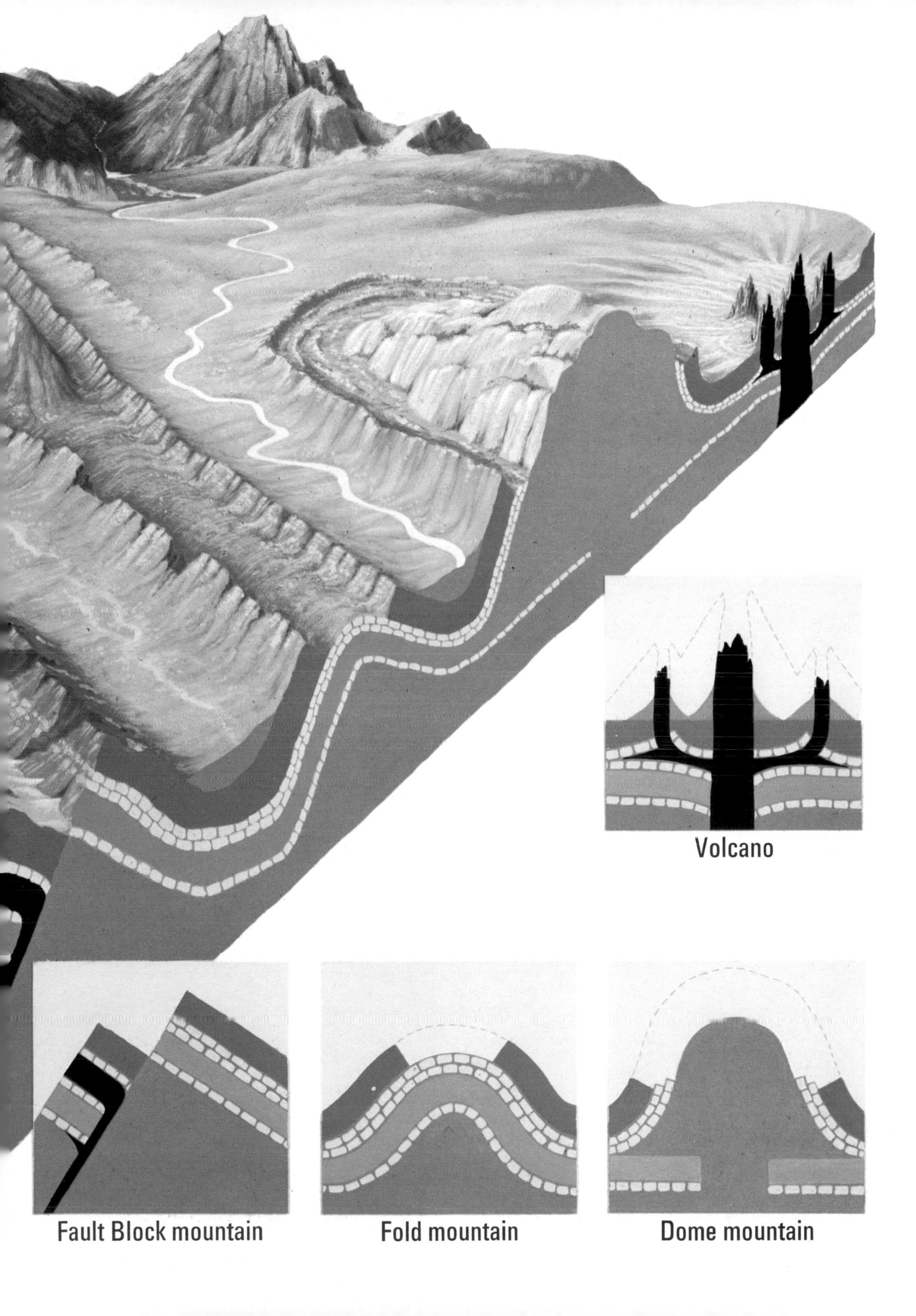
Volcano
Fault Block mountain
Fold mountain
Dome mountain

Mount Tarawera in New Zealand's North Island. Split asunder by an immense volcanic explosion in 1886, it reminds us that New Zealand stands astride an unstable section of the earth's crust.

positions. It has also ensured that individual plates grow, "shrink," and change shape.

Practically all *orogeny* (mountain building) occurs where two plates abut. If they are moving apart, the gap left between them is filled by new crustal material welling up from the molten mantle below. At present this happens almost entirely on the ocean floor, where the new material solidifies and becomes part of the plates on either side of the gap, forming an irregular ridge. The mid-Atlantic Ridge marks the line of divergence between the massive American plate on the one hand and the African and Eurasian plates on the other. Its growth is a record of the

Two mountain formations created by rifts in the earth's crust. Above: barren, clinker-surfaced Ascension Island, built by eruptions from the suboceanic Mid-Atlantic Ridge. Left: beyond these gazelles rises one of the mountain walls that line East Africa's Great Rift Valley, a result of crustal slumping.

slow, westward swing of the Americas. Volcanic activity along the ridge has produced many submarine mountains, some of them tall enough to pierce the ocean surface as islands. The Azores and Cape Verde Islands, Ascension Island, and Tristan da Cunha are mountains of the complex Atlantic ridge system. Iceland, too, is part of the ridge. Similar but more complex ridges wrinkle the floor of the southern Indian Ocean between the Indo-Australian and Antarctic plates; others seam the southern and eastern Pacific Ocean, at the trailing edge of the huge Pacific plate.

A similar rift occurs in Africa. Here, though, tensions have formed within a single continental plate, not between separate plates. The result is a massive crustal slumping rather than the growth of a boundary ridge. This created the Great Rift Valley system extending from the Jordan Valley in southwest Asia through the Red Sea and down East Africa to Mozambique. The African rifts probably began with the breakup of Pangaea some 200 million years ago. They are still active, with hot springs, *fumaroles* (holes that emit hot gases and vapors), and about 30 live volcanoes. Some geologists believe that these rifts will in time open farther to form a new ocean basin. Others see them as the product of earlier tensions, now permanently quiescent.

Plates that converge give rise to completely different boundary structures. It is this process of convergence that has thrust up many of the world's great mountain chains. But convergence products take several forms that largely depend on the materials involved. Where two edges of suboceanic crustal material are forced together, one appears always to be pressed down beneath the other, descending at a steep angle into the earth's interior. Warmed by friction and by the intense heat of surrounding deep-seated rocks, the descending edge of the crustal plate melts.

Coral-fringed Raiatea is one of the many mountainous islands pushed up in the Pacific Ocean by volcanic action. The volcano is extinct, but the mountain often generates a cloud. Such a cloud is visible at sea long before its island comes in sight.

Most of its materials become part of the mantle (from which, many millions of years earlier, they were originally derived). Some may be forced upward through cracks in the overriding plate (which is usually buckled and distorted by pressure), covering the seabed with lava or building mountainous volcanic islands. The line where the two plates meet is marked on the seabed by a deep, curving trench, often flanked by an arc of volcanic islands along the rim of the static plate. Many earthquakes originate at the interface between the plates, often deep down.

Most of the world's major island arcs occur on the rim of the Pacific Ocean, around the leading edge of the Pacific plate. New Zealand, the Philippines, Carolines, Marianas, Ryukus, Japan, and the Kuril and Aleutian Islands are all mountainous and rise steeply from the sea with deep submarine trenches close at hand.

Different types of mountains may result when a continent-carrying plate suffers continuing losses from the oceanic crustal material making up its leading edge. When this occurs the continent itself gradually approaches and reaches the edge, to confront the neighboring plate beneath which its own oceanic crust has been vanishing. Because of its buoyancy, the continental material does not follow the crustal material down the side of the trench. Instead, the roles of the two plates are reversed: the continent overrides, and the edge of the other plate slides obliquely beneath it to destruction. But in advancing, the edge of the continental block buckles, producing fold mountains, upthrust blocks and faults.

Geophysicists believe that these events have happened all along the eastern shore of the Pacific Ocean, where both North and South America lie close to the leading edge of the westward-swinging American plate. Mountain building is especially active in the south, where substantial pressure over the last 100 million years, punctuated by short bursts of violent activity, has buckled and lifted the exposed western flank of South America and neighboring seabed to form the Andes. The complex chain of the Rocky Mountains and the Sierra Madre of Mexico were built up in much the same way, from rock squeezed by the thrust of the Pacific floor plate. While the Andes and the mountains of the Panama Isthmus are still rising, pressures on western North America seem to have slackened and the Rockies are comparatively peaceful.

Where two continental masses clash on adjacent plates, both abutting edges are folded and uplifted, together with any sea-floor sediments that had piled up between them as they approached each other. This appears to have happened along a broad belt of southern Europe and Asia, where the Eurasian plate clashed with the African and Indo-Australian plates, producing a continuous sweep of mountain ranges that include many of the world's highest and most spectacular peaks. The Atlas Mountains of North Africa, the Pyrenees and Sierras of Spain, the European Alps, and many lesser ranges of Europe and Asia Minor, the Hindu Kush, Karakoram, Himalaya, and the mountains of southern China, mainland Southeast Asia and Indonesia were all raised in this way, reaching a climax during the last 20–30 million years.

The mountain chains so far considered are all relatively new. They formed at the boundaries of existing plates, and many are still a-building. That they include some of the world's highest and most striking mountains is not really surprising; though persistent weathering has begun to shape their peaks, it has not yet had time to soften their jagged outlines and round off their contours. But scattered about the continents are also the stumps of much older mountains. Although many have been further uplifted in recent activity, their peaks have weathered longer and their profiles are generally lower and gentler than those of newer mountains. Like their modern counterparts, those ancient mountains were built at the edges of converging crustal plates, but at times when the whole pattern of plates and continents differed sharply from that of today. The mountains of Scotland, northern England, Ireland, Norway, east Greenland, and northern Newfoundland were formed originally during the Caledonian orogeny, which reached its climax almost 400 million years ago. The Appalachians, and the eroded stumps of mountains in Cornwall, south Wales, Spain, France, and West Africa, belong to the younger Hercynian orogeny of 250–300 million years ago. The Ural Mountains, the massive ranges of Mongolia and northern China, the Great Dividing Range of eastern Australia, the Karroos of southern Africa, and many of the ice-bound mountains of Antarctica owe their structure to these old systems of folding. Traces of even older orogenies, dating from well over 1000 million years ago, can be found in the continental heartlands.

We have seen that modern mountains are the products of long and complex spells of crustal unrest. Similarly, slow, complex evolutionary processes gave rise to today's mountain plants and animals. Before the break-up of Laurasia and Gondwanaland, and for long afterward, the world was flatter and warmer than at present. After the Hercynian orogeny of Permian times there was at first little mountain building, and slow but inexorable weathering wore down the continental blocks to a mean height of 1000 feet or so, less than half their present elevation. Warm, shallow seas fringed with coral reefs flooded the continental margins. Luxuriant forests of ferns, horsetails, cycads, and conifers covered vast lowland plains, alternating with swamp and marshland. Amphibians and later on reptiles were the dominant land animals. Birds and small mammals appeared and began spreading widely as the continental blocks split and drifted away from each other. Although equatorial regions were probably no hotter than at present, subtropical conditions extended into far higher latitudes, and there were no polar ice caps. The mean temperature of the earth's surface was probably about 68°F, some 14°F higher than at present. High ground was cooler than the plains below, just as it is today. But few of the old mountain masses that survived could have supported permanent ice caps of their own, and there were probably no specialized mountain plants and animals anywhere.

Late in the Mesozoic era (which ended about 65 million years ago) flowering plants replaced ferns and conifers as the dominant plants of the earth. Dinosaurs and numerous other reptiles that had been evolving disappeared abruptly. By the start of the Tertiary period, about 65 million years ago, many familiar temperate and tropical trees had appeared and were forming widespread deciduous forests. Mammals, which had probably begun as insignificant insect-eaters, were now starting to diversify and dominate many terrestrial *food chains* (chains with links composed of eaters and eaten). The Atlantic Ocean, widening steadily, extended north to the forest-fringed Arctic basin. Antarctica and Australia, still joined together, were drifting southward across an expanding Indian Ocean, and the Indian subcontinent was drifting north and east toward its impact with the rest of Asia.

About this time the climate of the world began to change. Nobody is quite sure why this happened, but the shifting pattern of continental blocks could well have contributed. The most important change was a cooling of the extensive subtropical regions—at first slowly, then more

Glencoe in northwest Scotland. Aeons of erosion have whittled down the mountains overshadowing this famous pass. Scotland's present-day highlands are in fact a small reminder of the alpine peaks that stood here some 400 million years ago.

rapidly. Possibly the new arrangement of continents altered old patterns of ocean currents, largely blocking the flow of waterborne heat from equatorial to polar regions. Certainly by early mid-Tertiary times, 50–60 million years ago, the Antarctic continent replaced open water at the South Pole, and the north polar basin, once open, was almost landlocked. Whatever the cause, the polar regions cooled, and their chill spread.

Then began the cycle of earth movements that culminated in the Alpine orogeny, bringing high mountain ranges and intense cold to a world that had long forgotten both. From the middle of the Tertiary period onward, on every continent, thousands of square miles of low-lying land were slowly thrust upward into higher, thinner levels of atmosphere, to a climatic regime completely different from the luxuriant warmth below. For the first time in well over 100 million years, every continent gained a quota of high land that soon chilled sufficiently for annual snowfields and, eventually, permanent ice caps to develop. As ice and rain carved individual peaks and valleys, a range of new habitats emerged for local plants and animals to colonize, and new kinds of mountain life appeared on earth.

As more high land became snow covered, more solar energy was reflected back into space and lost to the world. Thus the earth's surface continued to cool at an increasing rate. Finally, some 2 to 5 million years ago, a threshold of cooling was passed, and permanent ice caps developed at the poles. First to form was the Antarctic ice cap. Then, in the Northern Hemisphere, ice spread over the lands and ultimately over the ocean surrounding the North Pole. Our planet had entered the Ice Age that many scientists believe still persists.

Photographs of melting mountain snowfields (a summer scene in Canada) and of Alaskan tundra (a cold-adapted type of vegetation) recall phases in the evolution of mountain life. Climatic change cleared once ice-clad slopes, letting plants and animals move in. As the glaciers retreated, tundra was the first type of vegetation to colonize the land left exposed by the vanishing ice.

Spectacular though these changes may seem, they occurred too slowly to affect the day-to-day lives of plants or animals. In the long term, though, we shall see that the changes profoundly modified the plants and animals through the slow processes of adaptation. Except where volcanoes erupted and lava cascaded, the most notable on-the-spot changes would have been gentle climatic fluctuations like those the earth is now experiencing. Even such variations are barely detectable in the life span of the longest-lived animals. Over many generations, however, mountain building and the Ice Age brought a range of cooler, drier, and on the whole less stable habitats into the warm, placid world of middle and late Tertiary times. Where rainfall dwindled, grasslands spread at the expense of forests. Conifers ousted deciduous trees in colder situations. Moorland and tundra vegetation skirted the permanent ice fields.

To organisms in the process of adapting for mountain life, the Ice Age presented new and formidable challenges. At least four times (perhaps many more) during the past one-and-a-half to two million years, spells of increasing cold caused mountain ice caps throughout the world to expand and descend to lower levels. Many sprawled outward to join newly formed ice sheets on nearby lowlands. Each so-called glacial period destroyed many mountain habitats and displaced all but the hardiest mountain plants and animals. During the relatively warm periods, and in the present post-glacial period (which will become the fourth interglacial period if the cold returns), ice caps have melted or shrunk, giving fresh chances for local organisms to explore, settle, and recolonize high ground. Today's mountain denizens therefore include both hardy, well adapted stock whose ancestors were mountaineers during the earlier interglacial periods, and relatively new recruits that have taken to mountain life since the end of the last glacial period, that is, during the past 20,000 years.

This process of recruitment has been well studied on the North American continent. Throughout most of the Tertiary period warm

temperate climates extended north to the shores of the Arctic Ocean. Intermittently during this period a broad highway of forested land linked America with Asia across what is now the Bering Strait. Semitropical animals including marsupials, primitive doglike carnivores, lemurs, lions, rhinoceroses, camels, and a host of insects, spiders, and other small creatures freely crossed this bridge between the continents. During the mountain-building movements of early and mid-Tertiary times many of these animals must have adapted for life in the newly emerging highlands, forming a pioneer mountain fauna. As northern latitudes continued to cool in middle and late Tertiary times, and successive orogenies raised mountains higher, a whole new group of cold-adapted animals crossed the so-called Bering Bridge from Asia and spread through the new pine forests and grasslands of North America. This wave of immigrants included primitive deer, lynx, bears, shrews, and lemminglike rodents; and their successors included hares, mammoths, wolverines, single-toed horses rather like modern forms, saber-toothed tigers, and many species of small rodents including the first meadow mice, squirrels, and pine voles. Some of these kinds of creature have found themselves at home in highland areas, augmenting the original mountain fauna and also competing with it.

Further cooling just before and during the Ice Age brought more than another 20 species of cold-adapted mammals from Asia to North America, many of them creatures that now occupy tundra and cold forest habitats both in the Arctic and in alpine zones farther south. Musk oxen and caribou, moose, bison, mountain goats and sheep, and several new species of deer were among the larger forms. Wolves, meadow voles, Arctic hares, Arctic foxes, tundra ground squirrels, and collared lemmings also appeared for the first time. Once on American soil, this final wave of immigrants dispersed south and east across the broad belt of tundra and coniferous forest that dominated central North America. During the four successive Ice Age glaciations, when ice sheets spread south from centers in the far north,

the belts of tundra and forest moved south too, and the cold-adapted species moved with them. When warmer conditions returned with each interglacial, the plant zones and their animals reoccupied the wastes left by the retreating ice.

Man himself invaded North America across the Bering Bridge during the final interglacial period, migrating south as the last great ice sheets advanced to the latitudes of New York, the Missouri River, and Vancouver. At the same time cold-adapted animals moved into the Rockies and the new ranges farther west, following corridors of alpine tundra and forest that flanked the mountain glaciers. During the last dispersal of the great northern ice sheets, which began some 20,000 years ago, many pockets of alpine vegetation and animals were left behind. As the climate warmed and the mountain glaciers retreated, these communities moved uphill to the cool upper slopes and peaks where we find them today.

What kinds of organisms have taken to living in the mountains? Although each continent has developed its own distinctive mountain communities, certain groups of plants and animals occur on mountains the world over. Among the plants, conifers seem especially well fitted to the cold, aridity, and high radiation levels of mountain slopes, replacing deciduous trees almost completely higher up below the timberline. Their resin protects conifers from disease, and snow tends to fall between their needlelike leaves instead of piling up and breaking branches. Each family of conifers has many mountain representatives. Cedars include the magnificent deodars found up to altitudes of 10,000 feet and more on the mountains of Afghanistan, Baluchistan, and Tibet; cedars of Lebanon flourish in the eastern Mediterranean region, including the dry Taurus and Antitaurus Mountains of southern Turkey; and Atlantic cedars once forested the damper slopes of the Atlas Mountains. Junipers have taken well to mountain living; *Juniper communis*, one of the commonest species, is widespread in the mountains of Europe and western Asia, and clothes parts of the Himalaya, Kamchatka, and mountains in North America. Several species that have developed a low, creeping form grow well close to the timberline in the Rockies and other high mountain areas. Spruces thrive at high altitudes, notably Engelmann's spruce of the Rocky Mountains and the hardy Himalayan spruce. Many species of pine cloak the shoulders of the world's mountains. In North America the white, limber, and ponderosa pines often reach timberline; Scots and stone (Arolla) pines of Europe flourish up to 6500 feet; and blue pines of the Himalaya form widespread forests at 10,000 feet and more. Firs and larches also grow well on mountains, often competing successfully with deciduous trees at lower levels.

In alpine meadows above and below the timberline appear the mountain representatives of many families of flowering plants. Most successful of all are the grasses, including the sedges (which flourish on wet ground) and the many species of mountain hair grasses, meadow grasses, fescues, and bents. These tend to be short-stemmed species, often hairy with partly rolled leaves, growing quickly in spring and

The bear cub probing this beaver pool (left) and the engaging tundra ground squirrel (right) derive from ancestors that entered North America from Asia by the now drowned Bering land bridge. As they fanned out across the continent, some kinds of creature moved uphill, giving rise to "native" mountain stocks.

maturing within the mountains' short summer season. They provide a favorable microhabitat for the more colorful flowering herbs that grace the meadows each spring and summer. Although dozens of families of flowering plants occur in mountain meadows throughout the world, a mere dozen or so provide many of the most prominent species. Why these particular families flourish especially well upon mountains is not clear. Some also thrive on the tundra, and the group as a whole has clearly adapted to a wide range of ecological conditions that embrace the alpine environment. These families and their best-known members include the Gentianaceae (gentians, with brilliant blue and pink flowers); Primulaceae (primroses, primulas, and oxlips); Ranunculaceae (alpine buttercups and anemones); Caryophyllaceae (carnations and pinks); Violaceae (violets and violas); Crassulaceae (fleshy stonecrops and houseleeks); Saxifragaceae (saxifrages); Ericaceae (lings, heathers, bilberries, and rhododendrons); and Compositae (including a wide variety of mountain daises and dandelions). High above the timberline conditions get harder, and flowering plants scarcer. Here they are largely replaced by mosses and lichens—ancient, primitive forms resistant to wind, cold, or drought, and surviving in tough corners the world over.

Although an enormous range of tiny animals lives in mountain soils and crevices, the most noticeable creatures—like the most obvious plants—belong to relatively few groups. Insects are especially numerous and diverse. By far the most generally successful animals on earth, it is not surprising to find them flourishing on the mountaintops. Many species and species-groups are widespread: closely related groups of beetles, springtails, and flies live on mountain slopes as far apart as the Alps, the Atlas Mountains and the volcanoes of East Africa, testifying both to their toughness and to their successful dispersal techniques. Spiders and mites are also worldwide mountaineers; with springtails some occur in the Himalaya well above 16,000 feet. Updrafts often carry butterflies and moths high among the mountains, and several species always live 5000–6000 feet up; Apollo butterflies of the Himalaya seemingly breed at more than twice that altitude. Mountain grasshoppers (generally wingless, incidentally) occur in many temperate regions, feeding avidly on mountain grasses and laying eggs that overwinter in soil. Snails, too,

Above: Apollo butterfly, a North American and Eurasian insect that breeds 12,000 feet up on Himalayan slopes. This specimen was appropriately photographed on edelweiss, a flowering alpine plant of high mountains in Eurasia and South America.

Below: boldly patterned asp vipers, a poisonous type of European snake that favors limestone hills and mountains. Asps may attain an altitude of 10,000 feet, but most live at much lower levels. They occur in the Pyrenees, Alps, and Apennines.

thrive on mountains. Although they rely on damp conditions for active life, they can retreat into their shells for long periods of drought and food shortage, and so survive in difficult climates.

Among vertebrates, both amphibians and reptiles generally cope worse with mountain conditions than the smaller cold-blooded creatures that we have just described. But many species in different parts of the world have climbed from lowland marshes and forests to live on the mountain slopes. Frogs and salamanders favor the damper zones of the high forests. Chameleons, skinks, and vipers inhabit areas above the forest edge. Asp vipers range up to 10,000 feet in the alpine mountains of Europe, skinks up to 13,000 feet in East Africa.

Because birds and mammals maintain constant body temperatures they can survive a wide range of environmental conditions, and both are well represented on mountain slopes. Many mountain birds are no more than daily or seasonal visitors, flying in to take advantage of locally abundant food. Some stay long enough to breed and molt, returning to the lowlands when the weather turns harsh. These birds are usually indistinguishable from lowland stocks. Others are true mountain birds, derived from lowland stocks but distinct enough in color or song to comprise separate species or subspecies. Several slightly different but closely related forms may occupy neighboring peaks of a mountain range, or neighboring ranges, while clearly owing their origins to a common lowland stock. Thus many races of Rocky Mountain water pipits have sprung from the same lowland stock as the water pipits of the New World Arctic tundra. Similarly the water pipits of southern Europe's mountains are cousins to west-coast Europe's rock pipits. Larger mountain birds tend to be cosmopolitan. Golden eagles and lammergeiers range widely across Europe, Asia, and northern Africa. Individuals from southern Spain and the Alps seem virtually identical with birds of the same species from Algeria and the Himalaya. Here is a hint that similar ecological conditions on isolated mountain "islands" beckon these large birds freely from peak to peak, so that even stocks that are normally apart may meet and interbreed.

Less mobile than birds, mammals have nevertheless adapted well to mountain life. Rodents abound on the uplands of every continent except Australia and Antarctica. Like insects, they are a versatile cosmopolitan group, adapting quickly

Lammergeier at its inaccessible precipice nesting site. This largest Old World raptor, a carrion eater, haunts high ranges from Spain east to north China and south to South Africa.

to changing conditions and taking to high, steep slopes as successfully as they have taken to polar tundra, deserts, atolls, and rain forests. Small mountain rodents, like small birds, are usually the close kin of nearby lowland species. Often several closely related species occupy neighboring zones. For instance, on the western slopes of the Californian Sierra Nevada, three related species of chipmunk—Merriam's, lodgepole, and alpine—occupy successively higher zones of chaparral, mountain forest, and tundra. Some of the larger rodents—marmots, for example—are more widespread, related species breeding as far apart as the Rockies and Himalaya. The dozen or more species of pika are found on mountaintops in North America, central Europe and eastern Asia. So broad a distribution can mean only that they were once widespread as a lowland species, but retreated to the mountains when forest replaced tundra in the temperate lowlands. Similarly the large blue or varying hares of Europe and North America, restricted to polar tundra and isolated moors and mountains farther south, were probably once joined in a single circumpolar population.

Most characteristic of all mountain mammals are the sheep, goats, and other horned ruminants that live among crags and pastures near the snow line in Eurasia, North America, and North Africa. These too are probably relics of once extensive lowland populations that retreated to the highlands, and eventually to the mountaintops, to escape competition and predation—possibly even the hunting of early man. Some species occur only in one area. They include the Rocky Mountain goat of the central and northern Rockies, the Dall sheep of Alaska and the Yukon, and the argali of the Tibetan plateau. Other species, such as the ibex, are distributed widely, with many geographical subspecies or races. Predatory mammals have not been slow to follow their prey species into the mountains: weasels, stoats, wolves, foxes, leopards, tigers, and many more carnivores usually have their base in the forests, but follow their prey far out over the tundra in summer.

Viewing a high mountain from far away, you can often see that base, middle, and peak must support different kinds of life. In some mountains the base is forested, the midriff grassy and the top bare. Elsewhere an entire mountain may be forested, but pale and dark horizontal bands indicate zoned tree communities. Although details vary geographically, practically all mountain vegetation is zoned in some way and the steeper the mountain, the better the zones are defined. However, few mountains are entirely symmetrical, and the zonal pattern is usually disrupted by spurs, gullies, landslides, and other discontinuities introduced by untidy nature.

In climbing from one vegetation zone to another, we pass through a succession of plant and animal communities. On many (not all) mountains climbing takes us from richer to poorer conditions. The lower communities tend to feature dense plant cover, fertile soils, and a rich variety of plants and animals; higher communities have thin plant cover, poor soils, and plants and animals that are few in kind and number. Mountain life, it seems, is organized in layers, with each layer a community, and the poorest communities at the highest levels.

This general rule gives us a basis for comparing the communities of different mountain ranges. In the chapters that follow we shall take a close look at a wealth of mountain communities—first those of North America and Eurasia, then those of Africa and the Southern Hemisphere. On our brief but far-ranging trip we shall meet some of the world's most hardy and enterprising plants and animals in the splendor of their upland settings.

Broadleaf and coniferous trees in unusual competition below the bald crags of the Jôf del Montasio in northeast Italy. In general, different plant types evolve as separate mountain communities forming zones that ring a mountainside at different levels.

Eastern North America

Aerial view of snow-dusted crests in Tennessee's Unaka Range, part of the Blue Ridge Mountains that form the southeastern end of the Appalachians. Like most of North America's eastern mountains the Unakas are wooded right up to their summits.

North America is roughly twice the size of Europe, covering about one sixth of the world's land surface. From Arctic shore in the north to tropical Caribbean coast in the south it spans 60 degrees of latitude; from Alaska in the west to Nova Scotia in the east is almost one third of the way around the world. Its vastness embraces almost every kind of climate, from the extreme cold of polar mountains to the extreme heat of tropical deserts, and almost every type of plant association except tropical rain forest. The mean elevation of North America is about 2000 feet. Like everything else on the continent, its mountains are built on a massive scale. Dozens of peaks exceed 10,000 feet and there are huge tracts of rolling upland above 6000 feet.

Two major systems of mountains cross North America. In the east the Laurentian Highlands and Appalachian Mountains form a continuous line from Labrador in the northeast to Alabama in the southwest. In the west stands the Western Cordillera—a complex system of roughly parallel ranges running mainly along a north-south axis and including the Alaskan highlands, Rockies, Pacific coastal ranges, and the high sierras of California and Mexico. Many smaller groups of mountains dot the continent. Except on the prairies, the northern tundra, and the flat lands of the southeast, few North Americans live far from mountainous country, and most are enthusiastic about the highlands that are part of their national heritage. In the past American Indians respected the spirits that they believed inhabited mountains; modern Americans take time off whenever they can to visit and pay their respects to the mountains themselves. In consequence, North America has an enviable array of state and federal parks, reserves, and other places of outstanding scenic beauty centered on mountain ranges or individual peaks. In such sanctuaries wildlife flourishes and is protected by sensible, informed legislation. Millions of visitors each year see something of mountain wildlife, if only through the windshields of their automobiles, and they pay handsomely to preserve it through state and federal taxation.

Let us now examine the mountains and mountain wildlife of eastern North America (saving the western ranges for our next chapter). The eastern mountains were formed by compression, folding, and uplift long before the Western Cordillera, and carry the marks of their greater age. They are on the whole much lower and more rounded than the mountains of western America, because more worn down by erosion. Although the most recent upheavals affecting the eastern mountains have occurred in the last 30 million years, their basic structure dates back some 250 million years, and mountains have stood where they stand for the whole of that period. For many millions of years they formed the eastern flank of a huge ocean, occupying the center of what is now North America. During the last million years many of the northeastern mountains suffered heavy and repeated glaciation. But those of the southeast escaped; though some were ice-capped, many stood high and dry to the south of the great Laurentian ice sheets, and retained at least part of their pre-glacial wildlife throughout even the coldest periods.

For descriptive purposes we may conveniently divide the eastern mountains into two groups: the Adirondacks and Appalachians south of the Saint Lawrence River, and the Laurentian Highlands to the north—though structurally speaking the Adirondacks form the Laurentian Highlands' southern extension.

The Adirondacks and Appalachians are relatively gentle, undramatic uplands. The Appalachians are part of an extensive system of fold mountains, which curve in a great arc from Newfoundland in the northeast to Georgia in the southwest. They are remnants of a complex block of sedimentary and *metamorphosed* rocks (rocks altered by great heat or pressure) that were thrown into curved folds over 250 million years ago, when Europe and North America were joined. Erosion over many millions of years has reduced these folds to long, narrow ridges of hard sandstones and shales, alternating with deeply incised bands of limestone—formations adding much to the region's variety of scenery and natural beauty. River dissection has cut deep channels across the ridges, breaking them into long, low mountain walls. In the north the mountains of central and southern Newfound-

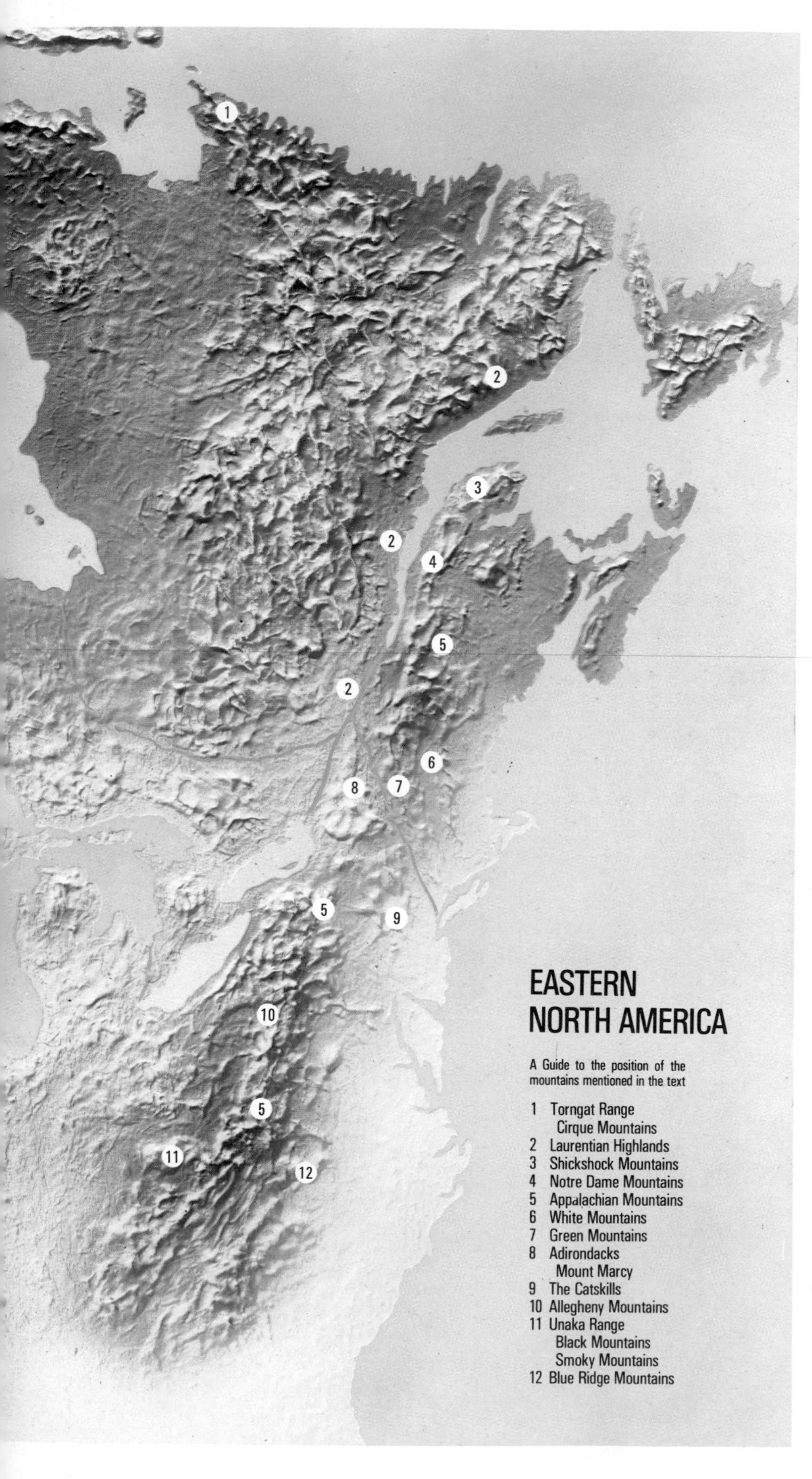

Mountains of eastern North America. The two main units are the Laurentian Highlands in the north, and the Adirondacks and Appalachians, which create a system south of the Saint Lawrence River. As a whole, the eastern mountains have low, gentle profiles—the product of many million years of erosion. Thus, except in the colder north, high-altitude life zones are absent or not well developed.

River otter (below) and bobcat (below right) are among predatory mammals that thrive in the northern Appalachians—though by no means exclusively there. Bobcats, for instance, live in forests or deserts between southern Canada and southern Mexico. This particular bobcat family plays in a Southwestern setting.

land, the Notre Dame and Shickshock Mountains of New Brunswick, the Green Mountains of Vermont, and the White Mountains of New Hampshire are part of the Appalachian system, with peaks between 4200 and 6300 feet above sea level. Farther south the highlands of New York State and New Jersey, the Blue Ridge Mountains of Virginia, and the Allegheny Mountains of Pennsylvania and West Virginia are also included. The Catskills are of slightly different structure and stand on their own. The highest peaks of the Appalachians occur in the Black Mountains and Smoky Mountains of the Unaka Range, Tennessee, where some rise above 6000 feet. The Adirondack Mountains, which stand in an enclave of the Appalachians in northern New York State, are a compact group of ancient granitic mountains, rising to a general level of 4000–4300 feet. Mount Marcy, the highest peak, exceeds 5300 feet.

Both the Appalachians and the Adirondacks lie below the level of permanent ice, but snow covers much of them deeply in winter. Most of their upper slopes are heavily timbered, with forests clothing all but the highest tops and ridges. These are the last remnants of the vast virgin forests that occupied New England and the southeastern colonies until white settlers cleared them for agriculture, lumber, and fuel. Now the fertile valleys between the mountain blocks are intensively farmed, and there are high meadows of lush summer pasture among and above the forest. But large areas of the mountains remain relatively unspoilt and serve as immensely popular summer and winter vacation playgrounds for the prosperous and populous east-coast states. Damaged by more than two centuries of severe hunting, mountain wildlife is now largely protected for the benefit of the tourist, naturalist, and licensed hunter.

Many sections of the Appalachians and Adirondacks retain their original pattern of zoned plant communities, as we discover if we climb their slopes. Such a climb starts a few hundred feet above sea level, usually from grassy plains or arable land. Before man cleared them for farming, deciduous woodland covered these foothills. Thousands of acres of original forest still survive in gullies and small, unworkable patches that escaped the firebrand and ax. But most of today's low-level forest has grown up on farmland abandoned during the last century as agriculture locally became uneconomical. The dominant deciduous trees are American oaks, beech, birch, maple, sugar maple, and hickory. American chestnut and elm, once plentiful, have suffered devastation by blight but other species have closed the gap to provide colorful low-mountain forest of unmatched beauty.

Deciduous forests have a layered structure of their own, each layer containing its own subcommunities of organisms that respond in different ways to the annual cycle of the seasons. The dominant trees provide a dense canopy overhead in summer, and a network of branches and twigs, buds, and leaves, where many plants, insects and birds make their homes. Bushes form an understory about the trunks of the trees. Lower bushes, annual and perennial herbs and grasses cover the forest floor. Beneath lies a rich, damp soil, tangled with roots and fungal strands, crawling with animal life on a tiny scale, and augmented by a bounty of mineral-rich leaves and twigs fallen from above.

Through winter the forest appears almost dead. As a general rule, the higher the forest the longer the "dead" season lasts and the shorter is the season of growth and abundance that follows. Usually soil is the warmest winter habitat. Fallen leaves, often covered with snow, trap summer warmth in the ground allowing

Forests of deciduous trees such as sugar maple (seen here in brilliant fall colors) shape patterns of life on the lower slopes of the Appalachians, where man has left these alone. The trees themselves support one set of organisms, soil holds another—based on nutrients in leaves shed by the trees. Right: a carpet of leaves fallen from wintry trees on New York's Bear Mountain.

activity to continue in the soil long after much else is stilled. In spring the forest floor is the first layer to show signs of life. Mountain forests facing south have a special advantage, for the sun strikes the ground at a favorable angle, warming and drying it quickly. So the bulbs and perennials that have stored food over winter produce their leaves and flowers early, taking full advantage of spring sunshine while the trees are still bare. Late spring sees buds bursting and leaves spreading in the canopy and understory. Flowering trees and shrubs attract insects, which in turn feed migrant birds that arrive in enormous numbers to breed. Summer brings browsing mammals to the forests, where they escape hot sun and biting insects, to feed on the lush growth of the forest floor. In autumn the trees and shrubs provide berries, nuts, and a crop of insects that fatten the birds and sustain them in their post-breeding molt. Small mammals, too, feed well at this time, laying up stores of body fat and underground hoards of food for the winter. Leaf-fall sees many of the forest animals disappear. Some birds migrate south, leaving only hardy wintering stocks, and many mammals move out or go to ground at the onset of cold weather.

The upper boundary of broad-leaf forest wavers and is locally hard to detect. In the north and on north-facing slopes it tends to be low; in the

south exposure to winds and many more factors affect the limit at which this kind of woodland can survive. There is usually a narrow transitional zone in which (deciduous) birch and mountain ash mingle with (coniferous) hemlock and spruce. Then, as we climb higher, deciduous trees peter out and conifers take their place—their tough, narrow leaves, seed-bearing cones, and shallow roots adapting them better than most broad-leaved trees for the poor soils and prolonged—often intense—cold of the upper slopes. Black and white spruce, balsam fir, and red spruce dominate much of the eastern North America's coniferous forest. Tamarack, balsam fir, and red spruce grow well in wet, marshy patches of ground; and red spruce and Frazer balsam fir are among the hardy species that survive at the timberline.

Coniferous forests of the plains—especially those planted by man in dreary, profit-yielding blocks—are dark, rather lifeless places with a sterile, needle-strewn floor and an almost complete lack of ground vegetation. Mountain forests of conifers tend to be far more lively, with a richer variety of plants and animals. The chief reason for this difference is the amount of light penetrating the evergreen canopy. In mountain forests, undulating ground (which places the trees at different heights), the mixture of species and the presence of young, middle-aged, and old trees all help to break up the canopy and let light through. Natural clearings, browsing mammals and other factors increase the variety of habitats, so that many species of plants and animals can find homes in the mountain forest and build up its ecological communities. Some of these organisms are drawn from the broad-leaf forest zone below. But conifers provide an environment

that differs in subtle ways from that of the broad-leaf forest, and we see striking differences in flora and fauna as soon as we pass from one zone to the other.

The coniferous mountain forest is usually colder and wetter than the broad-leaf forest and often stands in mist and cloud with a damp, dimly lit atmosphere of its own. This discourages all but a few specialized shrubs and flowering plants, but favors the growth of algae, mosses, lichens, fungi, ferns, and other nonflowering species that thrive in shade. So coniferous forests usually lack an understory, except for young conifers, but the ground is often carpeted with green, shade-loving plants. Their soils are usually poor, often damp and ill-drained in spite of slope and undulations. Resin-impregnated and encased in hard cuticle, conifer needles resist decomposition and tend to collect in sodden heaps, rotting slowly and releasing their minerals at miserly rates. The soils remain immature, developing a *hard pan* (a layer of insoluble minerals just below the surface) that roots cannot penetrate. Fortunately, the algae, mosses, and lichens make few demands on soil. Indeed many live on the trunks and branches of the trees, and festoon rocky outcrops that afford them little more than anchorage and a place in the cloud-filtered sun.

Let us now take a closer look at the creatures of the forests that largely clothe the Appalachian and Adirondack mountains. Some animals have

Misty air softens the outlines of fir trees in a mountain forest. In the Appalachians and similar ranges conifers thrive above the deciduous zone, for these needle-leaved trees withstand cold, damp conditions better than the broad-leaved trees.

been lost from the lower slopes through hunting and other activities, especially felling, which has largely broken the once-continuous deciduous forest into isolated blocks. Creatures of the coniferous forest have suffered less from human interference. Growing in higher, cooler regions above the deciduous zone, the coniferous forest proved less attractive to man, and much remains fairly intact. But both kinds of forest are well endowed with wildlife. Some kinds of animals (insects and birds in particular) are adapted for living and feeding in only one forest zone; indeed certain insects inhabit only one form of tree or shrub. But it seldom pays an animal of the high mountain forest to become so specialized. Whereas small mammals keep to their territories, birds and large mammals tend to move freely between deciduous and coniferous forest, taking the different foods—buds, leaves, new shoots, fruit, seeds—as they come into season, and emerging to feed on the alpine grassland and tundra during the summer.

The forest floor is the home of many small, herbivorous rodents, some adept at climbing trees and leaping—even gliding—from branch to branch. Best known of all are the squirrels, which feed mainly by day and are easy to see as they run and jump among the lower branches. Red squirrels are widespread in the deciduous forests, but they also live among pine, spruce, and other conifers in the northern and central Appalachians. Eastern gray squirrels are more restricted to the broad-leaf trees. Eastern fox squirrels occur only in the central and southern Appalachians, where they inhabit both kinds of forest and largely feed on the ground. Eastern chipmunks are small squirrels, complete with bushy tails. They, too, feed mainly on the ground. Unlike arboreal squirrels, the chipmunks nest in burrows and may hibernate while snow covers the forest floor. The even smaller flying squirrels have a fold of skin along each flank. These folds act like parachutes that help the squirrels to make long gliding leaps from tree to tree. All squirrels eat mainly vegetable foods, including nuts, berries, seeds, bulbs, and fungi, but they also take eggs and insects in season. In hard weather they strip bark to feed on the soft tissues underneath, and sometimes damage growing shoots. However, their engaging habit of distributing and burying nuts, many of which are forgotten and left in the soil to germinate, keeps them in favor with all but the most tidy-minded foresters.

Smaller mammals of the forest floor include white-footed mice, deermice, and (in the far southern Appalachians) golden mice, with a scattering of red-backed, meadow, yellow-nosed, and pine voles. Woodland jumping mice are also to be found on the forested mountain slopes, especially in damper areas close to streams and bogs. Larger herbivores include the porcupine, present on high forested ground throughout the northern Appalachians, and beavers and woodchucks, which extend to the southern end of the range. Moose, wapiti, and several smaller species of deer browse and graze in the mountain forests, especially in glades of the deciduous zone; moose are seldom found far from water, and they like to eat lush lakeside vegetation.

Omnivorous animals of the mountain forests include black bears, striped skunks, raccoons, wild pigs, and occasional coyotes. All take a variety of foods including nuts, fruits, eggs, carrion, insects, honey, and even young shoots and leaves. The large predators of the region are lynx, bobcats, and red and gray foxes, which live mainly on small mammals but also take birds, insects, and fruit in season. These are mostly solitary animals, which roam over wide areas in search of their prey, especially in winter. Smaller carnivores of the mountain forests include martens, short-tailed weasels, and martenlike fishers, found mostly in the northern sections of the Appalachians, and least and long-tailed weasels, mink, and otters, which are generally more widespread. The weasels usually eat mammals ranging in size from mice to rabbits, and also take eggs and young birds from nests. Fishers prefer larger mammals and birds. Martens, otters, and mink find much of their food in ponds, lakes, and streams. Hairy-tailed moles and several species of shrew are also present, feeding mainly on insects and earthworms from the forest litter. Red, hoary, and many smaller species of bat roost among cliffs and in hollow trees of the forest, swooping out in the early evening to feed on flying insects. Birds are especially plentiful in the high forests; ruby-throated hummingbirds, flickers, flycatchers, jays, chickadees, nuthatches, creepers, wrens, and many small songbirds occur as summer visitors or as year-round residents, with several species of woodland owl as their chief predators.

South of the Laurentian Highlands few peaks rise above the timberline, though its level varies with local conditions of soil, rainfall, and exposure. In the southern Appalachians the relatively warm climate allows trees to survive to the mountaintops. But in the north the timberline descends closer to sea level. Although vegetation maps tend to show the timberline as a firm, undulating boundary, it may be hard to locate in the field. This is especially true in summer, when the upper edge of the forest is blurred by a broad zone of miniature, contorted trees—mostly firs and spruces. In winter these disappear below the deep snow on which their very survival depends. Then the full-sized trees, which stand clear of the snow, do indeed form a clear-cut boundary.

Above this timberline the stunted conifers crouch like misshapen dwarfs among the boulders as though waging a permanent war with the prevailing wind. Any impression that they are barely surviving is quite misleading; these are among the toughest organisms on the mountain. Their strong roots wind firmly among the rocks, providing an anchorage few of the forest trees could match. Despite their small size, many are older than their tall neighbors of the forest edge. But they have grown painfully slowly in a harsh environment, and the constraints imposed upon them are all too apparent in their form. In height they seldom exceed the average depth of winter snow. This is because the leading shoot from each tree's apex—the shoot that ought to produce fresh upward growth—tends to be nipped off by the intensely cold, dry winds of spring, while most of the tree remains protected in its relatively warm snowdrift. The trees are not just dwarfed, they are lopsided too, all their branches appearing to stream downwind like banners. This is because buds that attempt to shoot on the windward side of the trunks suffer damage, and only buds on the leeward side can grow fairly normally. Even so, the resulting branches tend to be ground-hugging, stocky and with unusually short distances between lateral shoots. The trunks themselves are often polished or etched by the wind-whipped snow of autumn and spring, which at low temperatures abrades like fine sand. In spite of all their apparent disabilities, these *krummholz* ("crooked wood") forms of conifer are viable trees, producing fertile seeds that help to maintain the transition belt as a permanent feature of mountain life.

Rising above the dwarf conifers is the true alpine zone, sometimes called alpine or mountain tundra because it superficially resembles the treeless lands of the far north. Seen from a distance, much of it seems to consist of naked rock

Plant-eating mammals abound in the eastern mountains. The beaver (seen above, building a shelter or "lodge") eats the bark of hardwood trees such as birch, maple, linden, and poplar. This is partly why beavers of the Appalachian Mountains live below the zone of coniferous forest. To dam the stream below their lodge, beavers fell trees with their teeth, eventually clearing a patch in the forest. The northern deermouse (right) thrives on the forest floor, where its food includes berries, fruits, leaves, nuts, and seeds. Deermice also eat insects.

Below: black-capped chickadee on a snowy birch stump. Several species of chickadee live in the forested eastern mountains, feeding on insects and seeds.

where nothing could live. In fact this is a peculiarly interesting and varied zone, harboring plenty of life. Parts of the alpine region are indeed bleak and bare, with only tiny patches of lichen and even tinier insects barely surviving the constant winds and paralyzing cold. Other parts are rich and colorful, with shrubs, well-cropped lawns and some of the alpine plant species that enliven suburban gardens.

The richest corners of the alpine zone are sheltered sunny areas where meltwater moistens soils in summer and snow protects them in winter. Here grasses and herbs collect, building up deep, rich soil among their roots. Dwarf trees—mountain pines and junipers, and dwarf species of willow and alder—take root and provide protection and shade. Rhododendrons, wild raspberry canes, and briar roses help to form windbreaks, and broad expanses of grassland make a stable, protective home for a variety of small, flowering perennial herbs. Where vegetation is plentiful it tends to create more, both by producing organic debris that collects in corners, holds moisture, and eventually turns to soil, and by distributing seeds and propagules (bits of stem and other fragments capable of taking root) that colonize new ground. So these regions of alpine garden can be extensive, covering the ground and decorating rock walls and crags with a tangle of vegetation. Buttercups and crowberries, Lapland rosebay and miniature azaleas, violets and brilliant mats of diapensia provide a wealth of color throughout spring, summer, and fall, often mixed with bright red, yellow, and orange lichens, vivid green mosses and multicolored foliage. As throughout summer the gardens are grazed by deer, hares, and smaller mammals, and forest birds move in to feed.

Higher and drier regions of the alpine zone are less favored, but may still be well covered with rough turf and a scattering of small flowering plants—buttercups, blueberries, snowbells, and potentillas. Damp, marshy patches are richest in species. But the brief growing season, its big daily temperate range, thin soils, and other hazards make this a highly selective environment that few plants or animals are equipped to deal with. Higher still, the ground cover becomes patchy; areas of moss, lichen, and coarse grasses survive, often centered on tufts of dwarf willow (trees that stand only a few inches high, but bear sizable catkins in early summer). Hardy insects and other small creatures live here.

Geographical origin as well as altitude affects the kind of plants and animals found in different levels and regions on mountains south of the Saint Lawrence River. As you might expect, the Adirondack and northern Appalachian mountains have recruited many of their plants and animals from the north; their tops are covered with typically Canadian trees, shrubs, grasses, and flowering herbs, inhabited by typically northern insects, birds, and mammals. Thus tundra red-backed voles and northern bog lemmings, moose, and lynx occur in some mountain habitats south of the Saint Lawrence River, and snowshoe hares extend far to the south through the New England states. In central and southern sections of the Appalachians, which missed severe glaciation, there is a much richer and more varied flora and fauna. Here, too, you may see Canadian species on the upper slopes, but many plants and animals are of southern origin. In the higher forests Canadian hemlocks grow

Left: krummholz *conifers, dwarfed and contorted by frost and wind, grow above true timberline level on some eastern mountains and on other exposed sites (as in this Alaskan scene).*

The snowshoe hare or white rabbit is a northern species that ranges south through the Appalachians into Virginia. Its hairs lose pigment in autumn, giving it a white winter camouflage.

side by side with the southern species red spruce and Frazer balsam fir. Brilliantly flowering rhododendrons and azaleas, also dogwoods and mountain laurels decorate the forest edge and alpine meadows. The transition forest zone of the southern Appalachians, which occupies a wide swathe of the lower and middle slopes, includes many New England trees—for example, beech, yellow birch, and sugar maple—together with southern yellowwood, Carolina silver bell, basswood, large-leafed magnolia, buckeye, tuliptree, hickory, and oak. Among numerous species of southern birds in the mountain forests are wild turkeys, cardinals, Carolina wrens, and Kentucky warblers. Then, too, Virginia opossums, white-tailed deer, bobcats, gray foxes, New England cottontail rabbits, and southern flying squirrels are all mammals of southern origin.

To see the full rigors of high mountain life in eastern North America we must travel north to the Laurentian Highlands where alpine zones are more extensive than in the Appalachians that we have so far been chiefly considering.

The Laurentian Highlands of eastern Canada are eroded remnants of ancient folded mountains, largely made of gneiss and granite. They may once have stood higher than the Rockies but are now reduced to gently undulating uplands. Severely planed by ice during the most recent glacial spell, they have remained ever since in a subarctic climate featuring long, bitterly cold winters and a short growing season. Now they are thinly plastered with poor, acid soils, which support coniferous forest in the south and tundra vegetation in the north where in places the mountain timberline descends to sea level and merges with the northern timberline where the Arctic tundra begins. The highest Laurentian mountains occur along the windswept coast of Labrador, where several ranges were tall enough to stand clear of the ice sheets. Cirque Mountain, in the Torngat Range, is a bare, sculptured peak over 5000 feet high, practically all of which is above the timberline. Much of the remaining high ground of the Laurentians lies between 1600 and 3300 feet.

The mountain forests of spruce, tamarack, and fir that thrive in this bleak region support many Arctic and subarctic birds and mammals, whose main centers of population lie to the north. Small mammals include boreal red-backed voles, yellow-nosed voles, and tree mice, which eat seeds and insects on the forest floor. Spruce grouse and other ground-feeding birds share this habitat and way of life. Red squirrels and (except in the far north) flying squirrels and eastern chipmunks are active in the trees, competing with white-winged and red crossbills and other seed-eating birds for pine, fir, and spruce seeds. On the mountain tundra, northern bog lemmings, Arctic and snowshoe hares, and rock and willow ptarmigan feed on new spring shoots, and migrant birds move up from the south for the summer flush of insects. Woodchucks (or ground hogs) burrow in soft, well-drained ground, and Arctic, masked, and water shrews find niches

Right: snow-clad Mount Washington, in New Hampshire, seen from Lost Pond. A treeless summit with some of the world's wildest weather surmounts the forested flanks of this highest peak in northeastern North America. Below: a cardinal resting. Cardinals are among many creatures of southern origin that now live in the southern and central Appalachians.

for themselves in both forests and tundra.

Although the harsh winter climate of the Laurentian high forest and tundra keeps out many species of small mammals, those that do live here are often numerous. After two or three seasons of mild weather when food is plentiful and breeding success high, their numbers may erupt extravagantly in local areas, giving a bonus to the semi-nomadic or wide-ranging predators that feed on them. These include short-tailed and least weasels, lynx, and both red and Arctic foxes. Great horned, short-eared, and northern hawk owls hunt the woodland, forest edge, and tundra, taking advantage of the long hours of summer daylight, and raising large broods when their prey is plentiful. Snowy owls breed on the Arctic tundra and normally winter north of the northern timberline. But they sometimes spread south into the Laurentians and beyond. This happens if a summer abundance of food has raised their numbers to levels that the breeding grounds cannot support during winter.

Moose and woodland caribou are the larger grazing and browsing mammals of the Laurentian Highlands. Moose are usually found in wetlands during the summer, feeding on the vegetation of ponds and lakes; cows and calves usually run together in small family groups of two or three, but bulls remain alone. The rut occurs in September or October after the bulls have acquired their new antlers, and calves are born in May and June. During the winter, moose feed in the woodlands, eating bark and young wood; thick, dense fur protects them from the extreme cold. Woodland caribou form small wandering herds that browse in the forests in winter, often climbing to higher levels in summer to feeding grounds on the alpine tundra. Both moose and caribou are followed by gray wolves, which single out calves and old or weak individuals for killing.

Clearly, in spite of man's interference, the eastern mountains preserve much of their old natural magic. But North American mountain life attains its most diverse and spectacular forms in the great rows of ranges that rise in the far west of the continent.

These pages show some of the animals found in the Laurentian Highlands of northeastern North America. Above left: red fox seizing a ground squirrel, possibly drowsy after emerging too soon from winter hibernation. Below left: two baby caribou. Unlike the widespread red fox, caribou live only in the north, subsisting on grasses, saplings, and, in winter, lichens. The spruce grouse (below), aptly nicknamed "fool hen" for its tameness, survives in remote coniferous forests to the timberline. It eats berries and spruce buds. The snowy owl (right) hunts on the open tundra, roaming as far south as the Laurentian Highlands only if its Arctic food supply runs low in winter.

Western North America

Hoisted by immense subterranean forces, then scoured by ice and water, this great rock mass glowers down on Lake Maligne in southwest Canada. Its vast, stepped precipices typify the craggy grandeur that makes the western ranges so spectacular.

Extending from below the Tropic of Cancer to well above the Arctic Circle, the broadly parallel ranges of North America's western mountains traverse Mexico, the United States, and Canada, ending in the cold wilderness of Alaska. Like the Andes, this so-called Western Cordillera system owes its origins to lateral pressure that developed as the American plate slid westward into collision with the Pacific plate. The most active period of building occurred about 60 million years ago, early in the Tertiary period. Both offshore sediments and continental rocks were involved, and for many millions of years the Rockies in particular were subject to violent uplift and buckling. Now the pressures on western North America are reduced; there is little relative movement between the two plates, and the western mountains are relatively stable.

Most of the ranges trend northwest-southeast. This is a complex system of mountains, however, containing remnants of several generations of mountain building and structures that include both regular folded ridges and highly irregular faulted and tilted blocks. Over 3700 miles long, between 90 and 900 miles wide, the western mountains contain a splendid variety of scenery and a wide selection of mountain wildlife.

The Rocky Mountains, which form the eastern flank of the cordillera, fall naturally into three sections—southern, central, and northern. The southern Rockies are concentrated in the western half of Colorado, extending south to Sante Fe in New Mexico. Much of this block consists of an ancient granite plateau, worn flat during the long millennia of Mesozoic erosion but later lifted bodily and exposed to fierce attack by wind, rain, and ice. Now the plateau is carved into several ranges of mountains with peaks of similar height—about 14,000 feet. The Medicine Bow Mountains and the Front and Sangre de Cristo ranges stand to the east, the Park and Sawatch ranges and the volcanic San Juan Mountains to the west. Between lie broad areas of grassy and forested upland called "parks," famed for their wandering herds of game animals. Mount Elbert, Mount Harvard, Pike's Peak, and Blanca Peak, all of more than 14,000 feet dominate the ranges, but Colorado State alone has over 50 named peaks of similar height, and more than 1000 exceeding 10,000 feet. West of the southern Rockies lies the Colorado Plateau, a region of colorful sandstone drained by the Colorado River and its tributaries.

Between the southern and central sections of the Rocky Mountains lies the Wyoming Basin, a huge, rolling tableland 6000–8000 feet high, dotted with volcanic ridges and *buttes* (isolated, steep-sided, flat-topped hills). Its rich soils once grew nutritious grasses and herbs that supported abundant game including deer, pronghorn antelope, and bison. Now it is grazing land for cattle and sheep. The basin's southwestern boundary is the Uinta Mountains of Utah, formed from a block of land that rose between faults in the earth's crust. This block-faulted group runs east and west and at its western end overlooks the Great Salt Lake and its surrounding desert.

The tall mountains of the Southern Rockies show a fascinating range of habitats. Based in subtropical desert, they rise through zones of temperate forest and grassland to cold alpine tundra. Systematic study of their slopes has intrigued generations of American naturalists and helped in the training of many fine ecologists. Some of the earliest studies—those of C. Hart Merriam and his colleagues late in the 19th century—were important in developing the concept of "life zones." This was an attempt to explain patterns of plant and animal distribution of the whole North American continent in terms of climatic gradients. From studies elsewhere in the United States and Canada, Merriam had recognized seven major zones of vegetation and associated wildlife. From south to north he listed *tropical, lower austral, upper austral, transitional, Canadian, Hudsonian* and *Arctic-alpine* zones. The most important characteristics of all but the first of these transcontinental zones seemed to him repeated in the successive zones he found on the mountains of the southern Rockies. Thus the same life zones occurred both vertically and horizontally, in response to similar vertical and horizontal climatic gradients. Though the concept of life zones has now been superseded by a more complex system of *biomes*, based on climate, soil types, and many other

WESTERN
NORTH AMERICA
A Guide to the position of the mountains mentioned in the text
1 Brooks Range
2 Alaska Range
Mount McKinley
Mount Wrangell
3 Ogilvie Mountains
4 Mackenzie Mountains
5 Saint Elias Range
Mount Logan
Mount Elias
6 Cassiar Mountains
7 Rocky Mountains
8 Mount Olympus
9 Mount Rainier
10 Cascade Range
11 Big Belt Mountains
Crazy Mountains
12 Bitterroot Mountains
13 California's Coast Range
14 Teton Mountains
15 Sierra Nevada
16 Wind River Range
17 Medicine Bow Mountains
18 Uinta Mountains
19 Pikes Peak
20 Park Range
21 Front Range
22 Sawatch Range
Mount Elbert
Mount Harvard
23 Sangre de Cristo Range
Blanca Peak
24 San Juan Mountains
25 San Francisco Peaks
26 Sierra Madre Occidental

ecological factors, Merriam's comparisons remain valid. It is still interesting to see how the flora and fauna change as we climb in the southern Rockies, and to compare these changes with those along a south-to-north transect between the southern United States and Canada.

Merriam's southernmost zone—the tropical—is absent altogether from the southern Rockies; indeed in North America only Florida and Mexico contain it. But several of the southern ranges of the Rockies rise from his second zone—the *lower austral*—which extends northward from the plains and uplands of Mexico to lap around their foothills. This is a desert zone, semitropical but arid, typified by the Painted Desert region of many-hued sandstones south of the Colorado River, and the desert areas of southeastern Arizona. Characteristic vegetation includes the tall saguaro cactuses, 30 feet and more high, which rise from a rough scrubland of sagebush, coarse grasses, and agaves, with mesquite and cottonwood trees providing scanty shade in perenially dry watercourses. This zone extends from about 3000 feet above sea level to 6000 feet or more, and it covers many of the foothills and lower mountains of the Rockies in the far south. On higher ground the air becomes damper and cooler. Between 6500 and 8000 feet in the west, slightly lower in the east, the vegetation shifts from desert to dry mountain forest. Small, twisted piñon pines and yellow-green junipers begin to appear, merging to form a low, open woodland with dense understory of sagebrush and scrub. This corresponds to Merriam's third—*upper austral*—zone. Although generally arid, it is green and far from infertile, and damp enough to support semicontinuous vegetation.

The lower and upper austral zones (also called the Sonoran zones, for Mexico's Sonora desert region) are surprisingly rich in animal life. Intense heat during the day and intense cold at night drive many of the smaller creatures to live underground in burrows and crevices, where temperatures are more tolerably even. During the heat of the day nothing moves in the desert, but there is intense activity in the evening and early morning. The smallest mammals, including ground squirrels, cactus mice, and other ground-living rodents, emerge to feed on seeds and new shoots. Seeds contain all the moisture that some rodents need. Other rodents perhaps get extra water by eating food mainly when it is damp with dew or has absorbed moisture from the cool night air. Collared lizards and several species of snake become active late in the day, moving over the still-warm ground in search of insects and other small prey. Lizards and snakes in turn fall prey to roadrunners. Other birds include the seed-eating Gambel's quail and the insect-eating flickers. All these birds feed mainly during the day, resting in shade during the late morning and afternoon heat. Several species of owl inhabit the desert region. Western screech owls and the engaging Whitney's elf owl nest in giant cactuses and in trees, often in holes abandoned by flickers. These owls feed mainly at night, quartering their territories in search of insects and small mammals.

North America's parallel western ranges follow the coast. They were thrust up as the crustal plate bearing North America collided with the plate forming the floor of the Pacific Ocean.

The juniper and piñon zone has several species of birds and small mammals that feed especially on the rich piñon seeds. These creatures include noisy flocks of piñon jays, and ground-living rock squirrels, piñon mice, brush mice, Uinta chipmunks, and white-tailed prairie dogs. Rock gullies and canyons in the piñon zone have a locally rich flora including ponderosa pine and deciduous maples, aspen, ash, willow, and alder trees, rooted in damp soils with a dense undergrowth of grasses, herbs, and shrubs. Havens of cool, moist air, these gullies harbor herds of mule deer and a variety of chipmunks, mice, and other small mammals in abundance.

Predatory weasels, foxes, and the raccoonlike cacomistles hunt among the undergrowth, and the hooded, hognose, and more commonplace striped and spotted skunks are occasionally found. Birds—many of them summer migrants—frequent the gullies. Orioles, phoebes, western and summer tanagers, redstarts, and flickers are all prominent, and black-chinned hummingbirds flit busily among flowers in the undergrowth.

Between 6500 and 10,000 feet in the southern Rockies stands the Colorado Plateau, a rolling tableland capped with volcanic rock and deeply divided by the Grand Canyon. South of the canyon is a subdivision of the Colorado Plateau called the Coconino Plateau, an area forested with ponderosa pines. This represents Merriam's fourth, *transitional*, zone—transitional between predominantly southern and predominantly northern plants and animals. The tall, vivid green pines form open, parklike stretches of forest, often with a sward of grasses and colorful

Bryce Canyon in southern Utah. Barren rock pinnacles overlook a distant valley of the upper austral life zone, where piñon pine and Utah juniper shelter mule deer and smaller mammals.

flowering herbs growing from the rich lava soil. Ponderosa forest is good grazing ground for mule deer, and small rodents—especially chipmunks and mice—are plentiful on the ground. But much of the life of this zone is in the treetops. Insect-feeding mountain chickadees (close kin of European titmice), white-throated swifts, and pigmy nuthatches are numerous in summer, and Steller's jays and various seed-eating birds feed on pine seeds as they ripen in the cones. Red squirrels and tassel-eared squirrels also feed actively in the trees, especially on the pine seeds and on new shoots as they appear in spring. Incidentally, tassel-eared squirrels live only on the Grand Canyon Plateau and form distinct subspecies on either side of the Grand Canyon.

Few species of southern origin reach the cooler regions above the transitional zone. Their places are taken by cold-adapted plants and animals that originated in cooler, temperate regions to the north, and here reach their southern limits. First and most prominent of the plant newcomers that the climber notices is the Douglas fir, which appears in small numbers in higher and cooler parts of the ponderosa forest and gradually replaces the pines altogether at higher levels. The San Francisco Peaks, which rise southeast of the Coconino Plateau, are heavily clad in Douglas firs about their base, at altitudes of 8000 to 10,000 feet. More typical of temperate rain forests in Canada and the northwestern United States, Douglas firs appropriately mark Merriam's fifth zone, which he called the *Canadian zone.* Throughout the southern Rockies they form narrow bands of dark, closely-knit forest rising from the ponderosa pine forest but nowhere growing far above 10,000 feet. The Kaibab Plateau, forming part of the northern rim of the Grand Canyon, stands 2300 feet higher than the Coconino Plateau on the other side. This difference in elevation is sufficient to bring it into the Canadian zone, and to clothe it with a mixed forest in which Douglas firs are prominent.

Throughout the Canadian zone Douglas firs predominate wherever soils are deep enough to give them a good foothold. On rocky outcrops and slopes their place is taken by limber pines, which are content with poorer and drier conditions, and the forest edges are enlivened by aspens, whose shimmering leaves provide welcome touches of pale green in spring and golden-yellow in autumn. Aspens thrive wherever the forest has suffered damage from fire, lumber operations, or gales. More open than the somber coniferous trees, grasses and herbs can grow freely below them, and aspen stands usually contain more wildlife than the neighboring forest.

This Canadian zone of the southern Rockies has many animals equally at home in the Canadian forests far to the north. They include black bears and cougars, coyotes, raccoons, and a variety of chipmunks, squirrels, and other small rodents. Red crossbills and pine siskins feed on pine seeds in the treetops, blue grouse scratch the ground for seeds and insects, and several species of owl—including the pigmy, long-eared, and saw-whet owls—hunt nocturnally for birds, insects, and small mammals.

Between 10,000 and 11,500 feet above sea level Douglas firs gradually give way to slender Engelmann spruce, which forms extensive forests on exposed ridges. Corkbark firs and bristlecone or foxtail pines stand among them, usually in poorer, rocky soils. This is the *Hudsonian* life zone, comparable in Merriam's scheme with the spruce forests of the far north. Like the northern forests, it reaches its limit at the timberline—the narrow band of stunted, weatherbeaten trees separating forest from tundra. The tundra corresponds with Merriam's *Arctic-alpine zone*, his seventh and final zone. This is a semi-desert region of coarse grasses, sedges, and low, wind-carved shrubs, ice-bound or snow-covered for most of the year. During the late spring thaw the ground becomes marshy, forming ill-drained meadows that support a tangle of mosses, rushes, and tiny dwarf willows and birches. Drier ground is brightened in spring by the tiny but brilliant flowers of alpine avens, saxifrages, potentillas, violas, and other perennial herbs, which typically grow from small cushions nestling in the rubble.

The spruce forest and tundra zones of the southernmost Rockies are poorly endowed with animal life. Few species of northern origin have managed to penetrate so far south, and few southern species have had time to become adapted to the harsh conditions of high forest and open mountaintops. Bighorn sheep occasionally move up in search of summer grazing, and wandering black bears and porcupines follow in their constant quest for berries, roots, and generally palatable browsing. Typical animal inhabitants are midges, blackflies, and other tiny winged insects, which live among the low vegetation. They breed in damp moss patches and summer pools, feeding as larvae on bacteria,

Pale yellow fall aspen leaves contrast with the blue-green of background conifers on a Colorado mountainside. This scene is typical of the Canadian life zone in the southern Rockies.

algae, and other tiny organisms. Adults survive only a few days, mating almost as soon as they emerge from pupae. In calm weather their mating swarms rise in clouds above the vegetation, but strong winds tend to keep them in shelter. Females can lay eggs without feeding, but their productivity is boosted by a meal of blood, usually from a bird or mammal. Often they feed voraciously on nestlings or young mammals, and some transmit virus diseases from host to host. Birds are by no means abundant on the bleak tops of these southern mountains, but dusky grouse, rosy finches, mountain pipits, and rock wrens are often to be seen walking or flying low over the tundra, scratching a meager but satisfactory living from seeds and insects.

Rocky Mountain National Park, in northern Colorado, typifies the northern section of the southern Rockies. Starting at an elevation of about 8000 feet, it lies in a broad valley of open grassland and ponderosa pine, which soon gives way to forests of Douglas fir and Engelmann spruce. The park is well-known for its herds of mule deer and wapiti, which feed in the lower forest and parkland in winter and climb to higher mountain pastures in summer. Streams passing through the forest are dammed by beavers, which form small lakes around their lodges for protection, and spend their summer months storing logs, branches, and twigs underwater for their winter food. Though the evidence of their activity is clearly visible, beavers themselves are usually nocturnal and seldom seen. Cougars or mountain lions—splendid russet and

Hunted and hunters of the northern part of the southern Rockies. Below: mule deer, showing "velvet," the soft, sensitive skin that covers antlers and is rubbed off when the antlers are fully grown. Below right: young pumas. Pumas hunt mammals ranging from deer to mice, first craftily stalking their prey, then pouncing on it. Pumas sometimes range as high as 14,000 feet above sea level.

white cats weighing up to 220 pounds—roam the higher forests, feeding on deer and smaller prey. Their heavily spotted kittens, are born in litters of three to six at any time of the year, and accompany their mother for several months to learn the skills of hunting. Bobcats, much smaller cats with stumpy tails and tufted ears, are also present but seldom seen. The park is famous for its birds, which include many summer migrants as well as resident forest owls, nuthatches, jays, chickadees, and hawks.

The alpine pastures of Colorado's Rocky Mountain National Park are enlivened by ground-living rodents, including golden-mantled ground squirrels, rockchucks (known also as yellow-bellied or golden-mantled marmots) and bushy-tailed wood rats. The ground squirrels are small, lively squirrels with orange-red faces, shoulders, and forearms and prominent white bands down the flanks. They normally feed on seeds, insects, eggs, and fruit, but make themselves popular with park visitors by taking crusts, sandwiches, sausages, and cigarette butts from campground tables. Pocket gophers, least chipmunks, tree mice, and long-tailed, mountain, and red-backed voles are less prominent rodents of the timberline and alpine pasture. Like the ground squirrels they feed on a wide variety of small food particles. Although each has its own ecological niche and particular way of life, there is considerable overlap in their feeding habits—all take whatever is palatable from a none-too-hospitable environment. Most remain mainly active through the winter, living in burrows under the snow and feeding partly on stored foods, partly on what they find in their burrowing and skirmishing. Only the chipmunks and squirrels hibernate for any length of time.

Rockchucks are larger rodents that live among loose rocks and boulders of scree slopes. This habitat they share with pikas—small rodentlike creatures more closely related to rabbits than to

rats or squirrels. Rockchucks feed on grasses throughout the spring and summer and—unusually for mountain animals—hibernate for four to five months from October through February or March. Pikas do not hibernate. In summer they collect and dry hay, stowing it away under the loose rocks of the scree for winter fodder. Both rockchucks and pikas have high, piping calls that carry surprisingly far on the mountain air, and may be heard when the animals themselves are invisible against their background of broken rock. Often these are warning calls, given when a predator approaches, but piping may also be a way of claiming territory, and thus indirectly an aid to spacing the animals out and preventing overcrowding. Rockchucks produce four or five young in early spring, usually some time before the ground has begun to thaw. The babies emerge from the warmth of the den during their fourth or fifth week, and probably breed for the first time in the following spring. Pikas produce their young later in the year than the rockchucks, often rearing one litter

Above: the high wall of the Teton Mountains, viewed from the Jackson Hole area, forms one of the most impressive sights of the central Rockies. Below: collared pika. This species lives in the northern Rockies, but a closely related form haunts many high scree slopes of the central and southern Rockies.

in May or June and a second in high summer.

The central section of the Rockies rises north of the Wyoming Basin, its main ranges extending westward into Idaho and north into Montana from a focal point in northwestern Wyoming. Like the southern Rockies, the peaks of these ranges are mainly made of granite and rise to a common height—usually between 11,500 and 13,000 feet. This brings them well above the timberline, and most are snow-capped for several months each year. Major groups of mountains include the Wind River Range, a scenically striking range with many peaks, forested canyons, and ridges, and small remnant glaciers high on its shaded slopes. Even more striking are the 10 granite peaks of the Teton Mountains, which rise steeply for 6500 feet or more above the valley of Jackson Hole. Their well-watered lower slopes support grasslands and forest, and the high plains at their feet provide winter grazing for wapiti and mule deer. The Bighorn Mountains, famed for their upland sheep, form a spectacular eastern outlier of the Rockies. Their highest peaks, rising to 13,000 feet, stand a clear 10,000 feet above the surrounding prairie. Once heavily glaciated, many still carry small remnants of permanent ice above their timberline.

The central and northern Rockies lack the southern mountains' low-altitude life zones, but at high levels contain more northern species than their southern counterparts. The central and northern Rockies are well endowed with national parks, for the forested mountains and parklands and their wealth of wildlife annually attract millions of summer vacationers.

Yellowstone National Park, which fills the corner where Idaho, Wyoming, and Montana meet, and Grand Teton National Park a few miles to the south, typify the central section of the Rocky Mountains. Yellowstone, the first national park to be set up in the United States, occupies a volcanic plateau standing at about 8000 feet. Much of it is forested with pine, fir, and spruce, but the park includes open rangelands where bison graze. The Yellowstone bison are a small, protected remnant of the vast herds that once widely roamed the high plains of the Western Cordillera. Massive creatures weighing up to a ton or more when fully grown, they are well protected by their long, shaggy fur against the extreme cold of the mountain winter, and can store huge quantities of fat to see them through the lean months. Moose, the world's largest deer, with broad antlers like huge spread-fingered

Winter in Yellowstone National Park. Hot water from geysers keeps the Firehole River ice-free in below-zero temperatures. Such open stretches of water attract deer and other mammals.

Bison bull bellows a challenge to a rival male in the Moiese National Bison Range, Montana. Remnants of once-huge bison herds persist in reserves of the central and northern Rockies.

hands, are plentiful in the wetter areas of the park, where they feed on pond vegetation and lush grasses. Wapiti, another species of large deer weighing up to 440 pounds, forage in small herds through the park, wintering on lower plains and forested hillsides but climbing high onto the alpine pastures for the summer. During their summer spell in the mountains the bulls tend to form separate groups away from the cows and their young. In the fall, when bulls have grown their antlers and polished them, the sexes meet again. Each mature bull tries to round up a small harem of cows. Successful bulls utter ear-shattering roars and fight fiercely to drive off rivals. Mule deer, which reach only half the weight of wapiti, are also widely distributed in the central Rockies and often seen in national parks. Like wapiti they migrate annually from summer to winter feeding areas. Many congregate in the valleys warmed by Yellowstone's geysers, which remain relatively snow-free throughout the winter. Pronghorn antelope are also present on the rangelands, grazing in small herds on the plains and hills.

Grand Teton National Park includes a broad river plain of sagebrush and deciduous forest, and the dramatic slopes of the Teton Mountains that tower above it. Pine, fir, aspen, and spruce dominate the forest, and the meadows and forest clearings are renowned for their colorful carpet of flowers in spring. Grand Teton, like Yellowstone, has a rich assembly of animals. Although wary of human visitors, many of these creatures allow themselves to be watched, photographed, and occasionally approached. Both of these national parks are famous for their begging

Larger mammals of the central Rockies include the herbivorous moose (the world's largest deer) and the omnivorous black and grizzly bears. Above: moose crossing a stream in Yellowstone National Park. Left: black bear seeking tidbits. Black bears (often brown in the western mountains) are bolder than grizzlies.

black bears that stand at the roadside with paws outstretched to receive buns and other tributes from passing cars. Black bears, which may be any color from dirty yellow-brown to blue-black, range widely across North America. In the western mountains they tend to be brown. On all fours many stand less than three feet high at the shoulders but can rear up to the height of a man to reach food. In spite of their bulk, the bears run and climb with agility. Normally solitary, they pair briefly in spring and then part company. A mother gives birth to young in winter, usually in a warm den deep in the forest where she may spend several weeks in a deeply drowsy state.

The cubs, tiny when born, grow rapidly on their mother's rich milk. Two is the usual number, and the cubs often forage with their mother for several months—sometimes up to a year—before reaching independence. Black bears eat practically anything including wild gooseberries, pine cones, honey, ants, birds' eggs, mice, and carrion. Their larger cousins, the brown or grizzly bears, also roam the Rockies and they too live in Yellowstone and Grand Teton national parks. Although as omnivorous as black bears, grizzlies are shier and prefer the wilderness to man-haunted highways and campgrounds.

The northern section of the Rockies begins west of the Yellowstone River Valley, continuing in the broad uplands and ranges of central Montana. These are picturesque mountains, but modest in comparison with those farther south. Many peaks of the Bitterroot, Big Belt, and Crazy mountains stand above 6500 feet, but few rise above 10,000 feet. North of the boundary with Canada the Rockies again become spectacular, with many peaks of 13,000 feet and above on the Alberta-British Columbia border. Here most peaks and high *cols* (saddle-shaped depressions between peaks) are permanently capped with ice,

and many of the high valleys contain substantial remnants of the glaciers that carved them. The Rocky Mountains continue through northwest Canada, forming the lower and more modest Cassiar Mountains of British Columbia and the snow-encrusted Mackenzie and Ogilvie mountains of the Yukon. The Rockies proper peter out in the Brooks Range, which runs westward across north Alaska to the Arctic Ocean.

West of the Yukon's Rockies run the geologically distinct Alaska and Saint Elias ranges. The Alaska Range gives rise to the Aleutian Peninsula and beyond the mainland coast to the Aleutian Islands arc. Several of North America's tallest peaks are gathered in this northwestern corner of the Western Cordillera. Mount McKinley, the highest peak of all, is a magnificent granite mass flanked by sedimentary rocks and heavily capped with ice. Rising to over 20,000 feet, it is said to stand higher above its surroundings than any other mountain in the world. Mount Logan, at 19,850 feet, and Mount Elias, 18,008 feet, are of similar build. Mount Wrangell, over 14,000 feet high, is an active volcano. The whole area is heavily glaciated, with active glaciers descending the valleys and calving into the sea at many points along the coast.

The highest crags of the northern and central Rockies are the undisputed domain of mountain goats, which stand about three feet high at the shoulder and weigh up to 330 pounds. Ranging south to Wyoming and north to the seaward mountains of southern Alaska, they winter close to the timberline and spread out over the alpine tundra in summer. Their long, shaggy fur coats of creamy white protect them from rain, snow, and the biting winds that blast their habitat throughout the year. Mountain goats live communally in small herds of 10 to 12. Their horns, small but sharp, are used in establishing hierarchies within the herds, and as weapons of defense against wolves, mountain lions, and other predators. The broad black hooves, with their hard, sharp edges, grip the rocks like the nailed edges of climbing boots, affording a sure footing on the steepest slopes.

Slightly below their range, though often extending to the higher crags in summer, are the two species of Rocky Mountain sheep—the Bighorn of the south and the Dall sheep of the north. Bighorns stand as high as mountain goats but lack their square-cut bulkiness. Their coats are thinner, and brown or gray with paler rumps. The massive horns are carried by both sexes. Those of the mature male grow from triangular bases that occupy the whole of the top of the skull, coiling back above the brow and forward at eye level to complete an elegant spiral on either side of the head. Females and young rams have smaller horns that curve back but do not achieve a full spiral. Bighorn sheep are wide-

Beached fragments of melting iceberg gleam like jewels in the weak sunshine of Alaska's Glacier Bay. Many such icebergs derive from the great valley glaciers that flow down to the sea from cold mountains in the north of the Western Cordillera.

spread on mountaintops throughout the Rockies northward to southern Canada. Herds of 20 to 30 animals are common, browsing and grazing on the alpine meadows. Rutting occurs in November, when the herds are seeking food at the forest edge. In spring they move up to higher ground, and the lambs are born in May or June as the new vegetation begins to appear. In summer the herds split, the rams tending to feed high on the upper slopes while the ewes and lambs remain on lusher pastures close to the timberline. Though still fairly plentiful throughout their range, bighorns have proved tempting and easy prey for sportsmen, and their herds are fewer and shier than they were. Dall sheep are similar to bighorns but generally white, more rarely gray or blue-black with pale shoulders, rump, and belly. These are restricted to northwestern Canada and Alaska, where they inhabit the high pastures and crags above the timberline.

The high mountain grasslands of the central and northern Rockies, and the poorer pastures of the tundra that lie beyond, are the home of several species of hare and rabbit. Snowshoe hares and whitetail jackrabbits are widespread across the plains and prairies, and both may be found high in the mountains all year round. Both are snow-white in winter, only their eyes

Below: white or Dall sheep (ewe and yearlings). Northernmost of North America's wild sheep, these hardy, nimble, goatlike creatures roam the mountains of Alaska and the western Yukon.

Rocks torn by ice from mountains and swept downhill by water create the foreground in this view of far-off Alaskan summits. One is Mount McKinley, highest peak in North America.

and the dark tips of their ears showing up against the snow on a dull day. In spring they change color to match their background, the hares turning dark brown and the jackrabbits a paler gray-brown. They breed in summer in open nests, and feed throughout the year on grasses, twigs, and the bark and buds of low-growing alpine shrubs. Snowshoe hares are well known for their varying numbers. In some years they are rare, in others plentiful, and some populations seem to peak every 10 to 12 years. This peaking is due to their high fertility, a relative lack of predators, and a run of good seasons. During hard seasons many young hares die before reaching reproductive age, and numbers remain low. In better years, when the thaw begins early and food is abundant, each pair may breed two or three times, producing as many as a dozen offspring by the end of the breeding period. Two or three such seasons, with mild winters between, may give rise to a population too big for the local wolves, foxes, coyotes, owls, and hawks to keep down. Balance is probably restored when the animals crowd each other out, and more predators move in from neighboring regions to feed on the surplus. Mountain cottontails are smaller rabbits of the western mountains, abundant in wooded areas and on the plains of short grass above the timberline. They nest in shallow burrows or depressions among the rocks and vegetation, feeding on grasses, twigs, and shoots.

The northern Rocky Mountains of Canada and Alaska, and Alaska's southern highlands, have their equivalents of the pikas and rockchucks farther south. The southern form of pika is replaced by the collared pika, and rockchucks

The hoary marmot lives on mountain tundra from Alaska south to Idaho. Body fat accumulated during summer months carries this small mammal safely through its five-month winter sleep.

give way to two other species of marmot—the slightly smaller woodchuck and larger hoary marmot. Woodchucks are at home in coniferous forest from eastern to western North America, and thrive in rocky gullies and ravines of the high northern forests of the Yukon and Alaska. Hoary marmots are animals of the open tundra. Above the timberline they live deep in the recesses of scree slopes and feed on the grasses and herbs of the alpine meadows. Both woodchucks and hoary marmots are large enough to build up substantial reserves of fat in late summer—enough fat to maintain them through a five months' winter hibernation.

In the mountains of the far northwest the timberline forms an indefinite zone at 4000 to 5000 feet or less. Above it lies a region of poor tundra soils, frozen hard in winter and waterlogged in summer, clothed in sour tussock grasses and a mixed association of dwarf willow, birch, and smaller shrubs. Despite its bleak appearance, this Arctic mountain tundra supports substantial populations of birds, including songbirds that twitter and call tunefully in early summer courtship. Savanna, tree, and golden-crowned sparrows, slate-colored juncos and horned larks breed and feed among the melting snows, taking advantage of the long polar days that give them almost unlimited hours for foraging. Black-capped chickadees, olive-backed thrushes, mountain bluebirds, and red-breasted nuthatches are among the many songbird species inhabiting the spruce forest just below the timberline, and magpies, Canada jays, and several kinds of woodpecker nest in spruce, aspen, and willow in the high valleys. Most of these birds migrate downhill in the fall, leaving their breeding areas after the first snowfalls of late August and early September. Golden eagles, bald eagles, marsh and sparrow hawks, and gyrfalcons all roam the northern mountains hunting small mammals and birds in open country, and great gray, horned, and hawk owls patrol the forests on similar missions.

The western mountains of the Western Cordillera form a double line of ranges that follow the coast like a streetcar line from Mexico to Alaska. The inner, eastern, ranges begin on the Mexican plateau as the Sierra Madre Occidental, continuing into California as the Sierra Nevada, a long, granitic fault escarpment deeply cut by rivers and glaciers. The high peaks of the Sierra Nevada approach 14,500 feet. Snowcapped for much of the year, they stand in striking contrast to the hot desert plains and fertile valleys of northern and central California. The escarpment of the Sierra Nevada faces east toward the Rockies across the Great Basin, a vast rolling plateau over 400 miles wide, which occupies most of the state of Nevada. The line of the Sierras continues northward into the Cascade Range, apparently following a long series of linear faults in the earth's crust. The southern Cascades of California are entirely volcanic; in Oregon and Washington the central and northern Cascades arise from a raised platform of crystalline bedrock, with a row of spectacular volcanic peaks of 10,000 feet and more dominating the skyline. Mount Rainier, 14,410 feet above sea level, is the highest of the Cascades, which continue into Canada as the Coast Mountains.

West of the inner coastal ranges, and separated from them by a narrow, down-faulted trench, stands the outer, western line of mountains. Lower than the inner range, these seldom exceed 6500 feet. Their western foothills form the coastal scenery of the United States' Pacific states. In the far south they form the bare peninsula of Baja California—"the naked arm of Mexico." Farther north they become California's Coast Range, and continue north as coastal mountains in Washington. In British Columbia and south Alaska the coastal mountains become a string of mountainous islands leading on to Alaska's Saint Elias and Alaska ranges—ecologically (though not structurally) connected to the northern Rockies.

The varied relief and climate of North America's "streetcar line" of extreme western ranges is reflected in their plants and animals. In the far south the sun-baked desert peninsula of Baja California bears a chain of arid mountains rising to 10,000 feet or more. Although the south receives frequent storms, the center and north are largely rainless, and cactus desert covers much of the lava soil. High land near the Californian border, and the coastal mountains of southern California itself receive enough winter rain to support patches of coniferous forest and *chaparral*—a dry, mixed vegetation including grasses, scrub, and low evergreen or deciduous trees with small, leathery leaves. But drought afflicts chaparral country from March through October, and few mammals can survive its hot, dry summers. Among the commonest mammals where moisture and shade are available are raccoons and gray foxes, golden-mantled and antelope squirrels, and Merriam chipmunks. The open grasslands support small, wandering populations of mule deer and pronghorn antelope, and bighorn sheep find their food among the high crags. Mountain lions range the high chaparral.

In central and northern California persistent onshore winds bring heavier rainfall throughout the year, and the coastal mountains are clothed in dense coniferous forest. The dominant trees are coastal redwoods—perhaps the world's largest living organisms—that reach heights of over 350 feet. Redwoods grow in mixed communities including Douglas fir, grand fir, and bay, sometimes with a colorful understory of lupines, ferns, rhododendrons, azaleas, and small conifers. Farther north Douglas fir, western red cedar, Sitka spruce, and giant arborvitae dominate the forests of the coastal mountains. In Washington's Olympic National Park, where annual rainfall exceeds 130 inches, the lower slopes of Mount Olympus are hidden in a dense, dripping forest of huge firs, cedars, and spruces to about 3300 feet. Above stands a zone of silver fir, white pine, and hemlock that gives way to an unusually lush mountain tundra at about 6000 feet. The park is famous for its elk, which are so big that some authorities consider them a separate species. They grow fat on the high mountain pastures in summer, retreating to the forest in winter. Mule deer, mountain goats, black bears, coyotes, and mountain lions are also found on the heights in summer, with marmots, pikas, and ground squirrels occupying their niches among the crags and upland meadows.

Behind the coastal mountains stand the Sierra Nevada of California and the Cascade Range of

These pages show four scenes from the western mountains of the Western Cordillera. Above: a weatherbeaten Sitka spruce upon the wet, windy, sea-facing slopes of Oregon's Coast Ranges. Below: dry chaparral vegetation in Arizona.

Above: young raccoon peering from its den in a hollow tree. Raccoons are among the most widely distributed North American mammals, but in southern California's mountain chaparral zone they avoid the driest areas. Below: cut off from the rain-bearing sea breezes, the bare, sun-bleached ridges of eastern California's Death Valley stand in a desert.

Oregon and Washington. The Sierra, a tilted, deeply dissected slab of granite over 370 miles long, rises above the central valley of California to almost 15,000 feet. Its western slopes receive moist Pacific air, and support forests of pine, cedar, fir, and redwood. The eastern slopes, facing inland across the Great Basin of Nevada, are much steeper and drier. Midway along the range is Yosemite National Park, centered on Yosemite Valley, a deep gash cut through the granite by glaciers of the Ice Age. The valley floor, some 2000–2300 feet above sea level, is clothed in mixed deciduous and conifer forest, dominated by ancient oaks. Above 3300 feet incense cedars, white fir, and ponderosa pine form the bulk of the forest. On the mountain slopes behind the valley, red fir and lodgepole pine mark Merriam's Canadian zone. Above about 9200 feet lies a narrow Hudsonian zone of mountain hemlock and western white pine, with a timberline of white-bark pine at about 11,500 feet. The Cascade Range, which continues the line of Sierra Nevada to the north, is generally wetter and colder. Mount Rainier, the highest peak of the range, is based in coniferous forest of Douglas firs, hemlocks, and other predominantly northern species that reach a timberline at about 4900 feet. Beyond lies a narrow zone of lush alpine meadow, carpeted with grasses, low flowering shrubs, and perennial herbs. This dies away abruptly at about 6600 feet, the lower limit of permanent snowfields.

The forest of the Sierra Nevada are well stocked with small mammals and birds, some also occupying lower levels in the coniferous forests of the Cascade Range. Sewellels, otherwise known as mountain beavers, till the damp ground beneath the trees. Rabbit-sized rodents like chunky, tailless rats, they dig elaborate burrow systems among the tree roots, and feed on ferns, shoots, and pine needles. Open ground among the Sierra forests has its populations of golden-mantled ground squirrels and brush mice, dusky footed wood rats, sierra pocket gophers, yellow-bellied marmots, and Townsend, least, and lodgepole chipmunks—all small rodents with mainly herbivorous tastes. They form the prey of a variety of weasel-like carnivores that reach their southern limit in the central Sierra and Rockies. Smallest of these carnivores are the short-tailed weasels, lithe little animals up to eight inches long, that hunt both in trees and on the ground, feeding mainly on mice and birds. American martens, longer and weighing over two pounds,

are tree-living predators of coniferous forests. They take the tree-living squirrels as well as smaller rodents and birds, and also eat insects and fruit. Fishers are larger still, weighing up to 11 or 13 pounds; though mainly predators of small mammals they are heavy enough to tackle beavers, porcupines, and even ailing deer. Largest of the groups are the wolverines. Mainly animals of the Arctic forest and tundra, they also live in the western Rockies south to the Canadian border, and there is an isolated population in the high forests and tundra of the Sierra Nevada. Wolverines are squat, bearlike animals with an awesome reputation for strength, courage, and skill in hunting. Certainly strong, with powerful jaw and forelimb muscles, they nonetheless owe their success largely to sheer adaptability; they are scavengers and opportunist feeders rather than hunters, with no enemies apart from man. Brown bears, mountain lions, and mule deer are among the larger mammals of the mountain forests and meadows of the far western mountains.

Throughout the Western Cordillera, ponds, lakes, streams, and rivers provide a range of habitats for specialized plants and animals. High mountain streams that begin as glacial melt-water, or runoff from the bare boulders of the uppermost alpine zone, are usually deficient in minerals and sustain only a poor flora and fauna. Lower down the slopes they are joined by spring-fed streams, which usually have a higher mineral content, and they gain more minerals and organic material in passing through the high meadows and woodlands. Despite low temperatures and ice and snow that may cover them for well over half the year, mountain streams of the upper forest zones can be rich in plant and animal life.

Many mountain streams are dammed from time to time by rockfalls and, low down, by the activities of beavers. The obstruction often produces a temporary pool, which lasts for a few seasons until the constant pressure of water undermines or shifts the blockage. Pools alter the character of a stream in many ways, and provide a foothold for new kinds of animals and plants in the mountain environment.

Mountain streams and pools of the northern Rockies were well endowed with fish before white settlers arrived, though many post-glacial streams and lakes lacked a wide variety of species and man thought them understocked. Fishing first for food and later for recreation, man has made it his business to introduce new kinds of fish and to redistribute native forms, so that practically every headwater now has its quota of trout and other catchable species. On the eastern slopes of the Canadian Rockies cut-throat, rainbow, and brook trout and mountain whitefish are dominant in the cold, clear head-water streams. These are mostly bottom-feeders, whose main food is probably insect larvae. Lake trout and Dolly Vardens (a brightly colored char) feed mostly on small fish, and ciscoes and golden-eye trout are mainly planktonic feeders. Mountain sucker fish, equipped with suckers derived from their fleshy lips, browse on the algal coating of stones and gravel. Mountain streams of the western Rockies are invaded each year by Pacific salmon, which breed in the higher reaches. Coho and quinnat salmon especially find their way to the highest streams, to lay their eggs in the stream-bed gravel and die exhausted.

Many species of birds live in and around the high torrents of the Rockies. Most spectacular is the vividly colored harlequin duck, which winters on the sea coast but flies inland in May to spend the summer high in the mountains. Pairs nest close to a stream, often on overhanging cliffs or trees, and dive for food in the turbulent water. They feed mainly on insect larvae, which they grub out from under stones in a constant fight against the current. Their young leap from the nest within hours of hatching, bobbing like corks on the water and gradually drifting downstream with their mother, learning to feed as they go. More remarkable is the tiny dipper or water ousel, a wrenlike bird that plunges into fast-flowing streams and trots along the bottom in search of food. Walking against the current, with wings and body inclined forward, it is partly held down by the force of water and fully in control. Dippers build cup-shaped nests of moss near waterfalls, where they are relatively free from predators. So long as its stream remains even partly open and accessible in winter, each dipper stays in its own territory throughout the year.

Thus the brooks, crags, and forests of the Western Cordillera carry an immense variety of plants and animals adapted for survival in mountain settings. Together, these organisms form one of the richest of all wildlife treasuries.

A scene near the Columbia River Gorge. Plants clothing these lush, ferny slopes get water from spray from the waterfalls (the dark veil of falling water in the background), and the abundant rain that falls in this area of the Cascade Range.

Eurasia

Europe and Asia together make up the world's largest landmass—the supercontinent of Eurasia. From the Iberian Peninsula in the west to the Chukotski Peninsula in the east it spans roughly half the globe; from the Arctic in the north to Indonesia in the south it extends over 90 degrees of latitude. Many of the world's highest and grandest peaks occur in the mountain ranges of this giant double continent.

Geographers draw the boundary between Europe and Asia along the line of the Ural Mountains, but this boundary has no biological reality. There is strong evidence that many European species originated in eastern Asia, and European organisms seem to have spread eastward from time to time. However, Asia south of the Himalaya Mountains *does* possess its own distinctive plants and animals.

Geologically, Eurasia chiefly consists of a complex northern block, crumpled by early phases of mountain building, but stable over a long period. During the past 70 million years intense pressures from the southwest have affected the landmass. Its northern mountains—including northern uplands in Europe, the Urals, and the massive ranges separating Siberia from Mongolia—are mainly remnants of Hercynian and older orogenies, unaffected or only moderately uplifted by later earth movements. Eurasia's southern mountains, by contrast, are young and in many cases still active, owing their present prominence to massive upheavals during the middle and late parts of the Tertiary period. The newer folded mountains of southern Europe, Turkey, and Iran are due to pressures developing between the southwestern edge of the Eurasian plate and the northern edge of the much smaller African plate. Farther east, the Eurasian plate has come into even more violent collision with the western half of the broad, crescentic Indo-Australian plate, giving rise to the Hindu Kush of Afghanistan, the Pamirs and Tien Shan of the southern USSR, the Karakoram and Himalaya

Winter in Czechoslovakia's Tatra Mountains—highest section of the central Carpathians. The Carpathians form part of the great Alpine-Himalayan mountain system that strikes across Eurasia from Spain in the extreme west to China in the east.

north of the Indian subcontinent, and folded mountains of southern China, Burma, Thailand, and Indonesia. The island arcs and recently formed mountains of eastern and northeastern Asia are due to pressure from the Pacific plate and possibly the American plate, producing uplift and folding on the continent itself and volcanic outbursts along its eastern edge.

Our tour of Eurasia's mountains and their life-forms begins in Europe. Europe's older mountains—those of Scandinavia, the northern British Isles, central Spain, south-central France, southern Germany, and western Czechoslovakia—are weathered remnants of folding and massive upheavals that long predated the movements which threw up the Alps. These remnants are made up of hard crystalline, metamorphic, or sedimentary rocks, which were severely planed and remodeled by glacial action and have since been largely weathered to rounded stumps. The younger mountains of southern Europe originated in the bed of a long, narrow sea—the Tethys Sea—which for many million years separated the supercontinents Laurasia and Gondwanaland created by the breaking up of Pangaea. Sand, mud, gravel, and limy oozes that had formed the floor of the Tethys Sea were lifted during the Alpine movements and thrown into massive folds, overlying the remnants of older mountain blocks. So Europe's southern mountains include both crags and massifs of ancient crystalline rocks and enormous areas of limestones, conglomerates (natural concretes), and other sediments. Both groups of rocks contribute to the varied topography and scenic beauty of ranges from Spain to Romania—some containing famous national parks.

Europe's most spectacular mountain system is the Alps, which stretch in an arc about 660 miles long from Mediterranean France to eastern Austria. The Alps are in fact several ranges of mountains: their crags and longitudinal folds alternating with deep, ice-scoured valleys to create a series of separate blocks. Many peaks rise above 10,000 feet; Mont Blanc, the highest mountain of the entire system ends in a sharp, ice-capped point at an altitude of 15,781 feet. The Alps still carry substantial glaciers and permanent or persistent snow lies above about

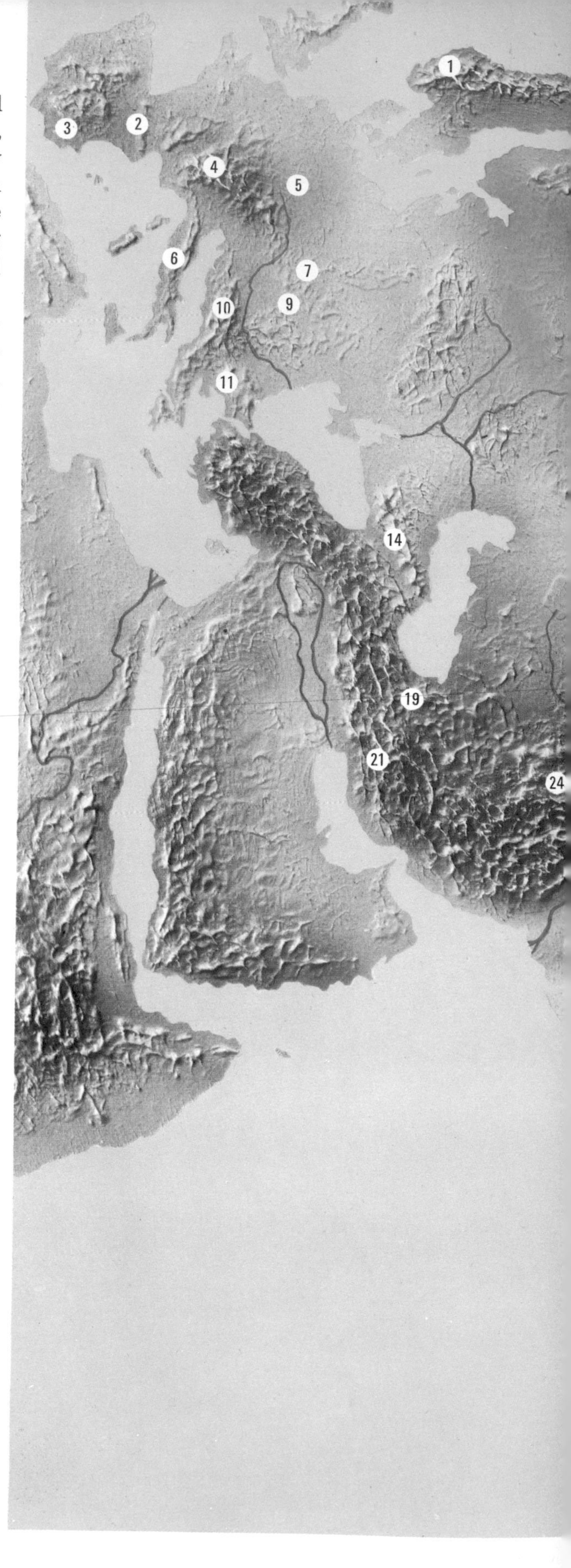

Major Eurasian mountain groups include the Alpine-Himalayan system from Spain to China; the Circumpacific system involving Japan and Indonesia; and more ancient systems.

EURASIA
A Guide to the position of the mountains mentioned in the text
1 Dovrefjell Mountains
2 Pyrenees
3 Sierra Nevada
4 Alps
Mont Blanc
5 Erz Gebirge
6 Apennine Mountains
7 Carpathians
8 Verkhoyansk Range
9 Transylvanian Alps
10 Dinaric Alps
11 Rhodope Mountains
12 Ural Mountains
13 Stanovoi Mountains
14 Caucasus Mountains
15 Sayan Mountains
16 Yablonovy Mountains
17 Altai Mountains
18 Tien Shan
19 Elburz Mountains
20 Pamirs
21 Zagros Mountains
22 Kunlun Shan
23 Karakoram Range
24 Hindu Kush
25 Tsinling Mountains
26 Himalaya
Outer Himalaya
Lesser Himalaya
Greater Himalaya
A Nanga Parbat
Kolahoi
Nanda Devi
Annapurna
B Everest
Makalu
Kanchenjunga
C Namcha Barwa

8000 feet on many northern slopes, above 10,000 feet on warmer southern faces. Deep valleys that wind between the mountains help to distribute summer warmth. People can thus live on large tracts of the Alps, even high above the timberline, for at least part of the year.

Most areas of the European Alps have been modified by man, usually to the detriment of wild plants and animals. The northern foothills, which rise in stages from the north European plains, were once covered with a temperate deciduous forest of oak, chestnut, ash, maple, and sycamore. Little of this forest remains, though modified patches persist among the fields and scrub that have extensively replaced it. Above about 1600 feet beech and fir predominate, with red firs replacing white as altitude increases. From about 2600 feet oaks and chestnuts become rare, except on sunny, sheltered faces. Above 3300 feet the forest is almost totally coniferous. White firs and beeches reach an upper limit at about 4900 feet, leaving red firs as the dominant species. On drier and more exposed slopes Scots pine predominates from about 3300 feet. Above 6500 feet the main northern massifs of the Alps carry a mixed coniferous forest in which larch and stone pine predominate over Scots pine and red and white firs. In favored situations this pattern extends to above 8000 feet, though the timberline for the northern and central Alps is generally 6500–7500 feet. The high coniferous forest has suffered less at the hands of man than the deciduous forests of the lower slopes. This is partly because the soils on which the high forest grows make poor farmland, and felling the trees would leave the ground open to damaging rainwash and avalanches.

The southern foothills, viewed from the Plain of Lombardy, were formerly covered by drought-resistant forests of evergreen oak and by *maquis*, shrub chiefly comprising aromatic evergreen shrubs. Such vegetation is typical of dry lands bordering the Mediterranean Sea. On the damper slopes above the foothills a deciduous forest of common oak and chestnut rose to 2600–3300 feet, higher again grew a mixed woodland of beech and red fir, capped by larch just below 5000 feet. Now large tracts of these southern foothills are farmed, with the help of irrigation provided by water from the alpine lakes. Olives, citrus fruits, and vines ripen in sheltered corners, and cereal crops and pine, spruce, and cedar plantations replace much of the native vegetation. Both agriculture and trees reach an upper limit at about 5000 feet.

Above the timberline grows a varying but generally narrow zone of shrubs, including dwarf pines, alders, alpine rhododendrons, and dwarf and savin junipers. Often these tough, wind-carved bushes provide shelter for smaller shrubs — heaths, (black-fruiting) bearberries, raspberries, and red elders—and windbreaks for the broad expanses of meadow grasses and flowering herbs that flourish at the same level. Alpine meadows extend throughout the Alps up to about 6000 feet where they are replaced by a patchy flora of tussock grasses, stone-crops, dwarf sedges, and the most hardy herbs of the mountain tundra—the stemless campions, mountain avens, saxifrages, and mountain gentians.

About one third of all the 200-or-so alpine plant species occur only in the upper zones; these are the true high-mountain forms. Remarkably enough, about 40 per cent of them live also on the polar tundra of the far north, separated from the Alps by over 1500 miles of lowland forest and plain. In general, species found in both polar and alpine habitats prefer damp situations. Several such species occur above the timberline in the Altai Mountains of central Asia, far to the east of the European Alps. Many species of insects, and a few kinds of birds and mammals have a similar discontinuous distribution in polar and alpine habitats. These stocks were evidently united over much of what is now a temperate zone when broad ice sheets covered the land, and have separated only during the last few thousand years as ice retreated, polar tundra crept north, and alpine tundra invaded the peaks.

Like plants, many creatures of the Alps have suffered at man's hands. Most seriously disturbed were beasts of the now dismembered forest of the lower slopes. Many of its large mammals have already vanished or are vanishing, notably brown bears, wolves, and wild cats; and wild boar and roe deer are rare. Smaller forest mammals fare rather better. Wood mice, edible dormice, garden dormice, shrews, moles, weasels, stoats, badgers, pine martens, and beech martens exploit both the forest remnants and the walls, orchards, and vineyards that have replaced much of the original woodland.

Wild creatures of the coniferous forest, alpine meadow, and tundra zones have suffered less than those at lower levels. But high up, man's domestic animals—notably sheep, cattle, and

Alpine ibex—one of Europe's several varieties of wild goat. Ibex range the rocky slopes of the Alps above the timberline, munching grass, lichen, moss, and the leaves of dwarf pines.

goats—compete with native species for grazing and browsing. Goats especially have left their mark on the Mediterranean slopes of the Alps, and their devastation is even more complete in other mountain forests around the Mediterranean. Again, the smaller kinds of wildlife are least affected, and the higher valleys and slopes of the Alps still retain their populations of snow voles, alpine marmots, alpine and southern water shrews, and blue hares. The highest crags are occupied by chamois, goatlike antelope that survive naturally despite many generations of severe hunting pressure. Alpine ibex—wild goats with big, curved horns—were reintroduced after local extermination and now live in small groups scattered widely throughout the Alps.

The higher slopes of the Alps have also kept their populations of amphibians, reptiles, and birds. Alpine ponds and marshes are the home of the tiny yellow-bellied toads, whose soft, bell-like croaking in late April is a welcome sign of spring. Alpine newts breed slightly later in similar situations. Fire and alpine salamanders inhabit damp crevices in mountain woodlands, and live in more open situations where damp moss and rocky corners provide protection. The black and yellow fire salamanders occur up to 3300 feet. Females give birth to 30 or more tadpolelike larvae in water early in spring, and the young remain in water for two to three months before metamorphosis. The black alpine salamanders are sometimes found on open ground 9800 feet above sea level. Female alpine salamanders keep their two young inside them until fully metamorphosed; the young then live as land animals from the moment of birth. Both species feed on slugs, caterpillars, earthworms, and insects during the summer, and hibernate in frost-free corners in winter. Aesculapian snakes, adders, asp vipers, and common or viviparous lizards are found in drier woodlands and sunny, open situations in the Alps; adders

The agile, wary chamois seeks grass, buds, twigs, moss, and lichen in high forests and on rocky slopes above—from Spain to eastern Asia Minor. Most troops are led by an old female.

and asp vipers occur up to 6600 feet, lizards up to 9900 feet in favored situations.

Alpine birds include the sparrow-sized alpine accentors, and snow finches, ring ouzels (9-inch-long dark birds with a white breast crescent) and alpine choughs (yellow-billed crows that nest among the rocks of the high grassland and tundra zones and feed on insects and seeds). Wallcreepers (grayish, crimson-winged birds with a floppy flight) are especially attracted to high cliffs above the timberline, where they nest in deep crevices and catch insects on the sunny rock faces. In winter they migrate to the valleys. Alpine swifts also nest among the high crags, catching their insects on the wing. They spend only a few weeks in the Alps each year, migrating south beyond the Mediterranean from fall to late spring. Three-toed woodpeckers and nutcrackers live in the high coniferous woodlands up to the limit of trees. Woodpeckers feed largely on insects and grubs that they catch on the tree trunks and dig out of the timber; nutcrackers have a stout, businesslike bill that extracts pine seeds from their cones. Rock partridges live among the dwarf junipers and rhododendrons above the timberline, feeding mostly on seeds, young shoots, and insects. Griffon vultures (restricted to the eastern Alps), golden eagles, and the rare short-toed eagles are the largest birds of prey; goshawks, harriers, kestrels, and other smaller raptors are often seen gliding or hovering in the ever-mobile mountain air.

Roughly central among the mountains of Europe, the European Alps are a focal point from which other ranges diverge, or appear to diverge.

To the east stand the Carpathians and Transylvanian Alps. The Carpathian Mountains, which arose at the same time as the Alps, bound the north and east of the Plain of Hungary in a wide arc that takes them within 100 miles of the Black Sea. At their most southerly point they turn sharply west through more than a right angle, forming the Transylvanian Alps of Romania. Although marginally longer than the Alps, the Carpathians are on the whole very much lower, with most of their high peaks below 6600 feet. But the northern Carpathians have many spectacular cliffs, crags, and gorges, and the southern Carpathians (including the Transylvanian Alps) display alpine scenery in miniature. Here several ridges reach 6500 feet, the highest Transylvanian peak rising to 8343 feet. The Carpathians have a more thoroughly continental climate than the European Alps: hotter in summer, colder in winter, and drier throughout the year. Although moderately glaciated during the Ice Age, they now lack permanent ice or snow. Most of their precipitation falls as summer rain, encouraging the growth of trees and mountain pastures to the tops of all but the highest summits.

Like the Alps, the Carpathians rise from agricultural, once-forested, plains. Above what remains of the low-level deciduous (oak and beech) forest is a narrow band of conifers, predominantly fir, pine, and spruce, that caps many of the lower peaks and ridges. From about 4600 feet in the northern Carpathians and 5600–5900 feet in the south, tall trees are replaced by a narrow zone of "elfin forest," featuring thickets of dwarf conifers. Alongside and above this formation the highest peaks support a thin cover of alpine grasses and herbs, which emerge from their mantle of snow for only a few weeks each year. West-northwest of the Carpathians lie older mountains—the Erz Gebirge and other ranges of western Czechoslovakia—and beyond these the heavily forested upland blocks of southern Germany and France. In many respects these resemble the northern Carpathians, though they are generally lower, with a few peaks above 4000 feet and little or no permanent snow.

Less heavily peopled than the Alps, the Carpathians and neighboring mountains have retained more of their native fauna. With fewer high peaks thrusting beyond the timberline they lack some of the alpine tundra species; alpine shrews, snow voles, and chamois are distributed patchily among the higher ranges, but there are no ibex or blue hares. Alpine marmots have been introduced into the Beskid region of the northern Carpathians, and mouflon—originally wild mountain sheep of Corsica and Sardinia—flourish well on certain high pastures in southern Poland and in Czechoslovakia. But the strongest populations of mountain wildlife are found in the forests, home of brown bears, wolves, wild boars, and red, roe, and fallow deer. Wildcats are common in remote areas, and the European lynx occurs locally in small numbers in the Carpathians. Farther south, in the southern Carpathians and Transylvanian Alps, lives the beautiful pardel or spotted lynx. Small mammals are well repre-

Above: handsomely spotted fire salamander from eastern Switzerland. This land-dwelling amphibian likes damp hillsides. Below: golden eagles hunt marmots and rabbits in the Alps, nesting locally in ranges across the Northern Hemisphere.

Shepherds graze their sheep to the very crests of many Pyrenean mountains. Here, summits tend to be lower, more softly rounded, and thus more richly clothed in grassland than those of the high, often jagged, snowy peaks of the European Alps.

sented, including the northern birch mouse (a lowland species in the north, but a mountain species in southern Europe) and the ubiquitous harvest mouse, field mouse, and common, ground, and bank voles. Predatory mountain birds include golden and lesser-spotted eagles, hobby and peregrine falcons, and Tengmalm's, Ural, and eagle owls. Vultures, too, occur in Romania's remarkable Retezat National Park, at the western end of the Transylvanian Alps.

Europe's extreme southern mountains are typically composed of folded sedimentary rocks in the alpine pattern. The Dinaric Alps, bordering the Adriatic Sea, form a chiefly limestone region of high forest and grassland broken and dissected by cliffs, crags, and deep river gorges. This is agriculturally poor country, difficult of access and thinly peopled. Largely for these reasons wildlife remains plentiful and varied. The Durmitor National Park of northern Montenegro, rising to over 6500 feet, is one of several in which tourists can see bears, chamois, deer, and other mountain animals. Snow voles are among the alpine species present, and a pale gray vole—Nehring's snow vole—is known only from the mountains of southern Yugoslavia. The fold mountains continue south into the Balkans, diverging to flank a triangular block of older rock—the Rhodope Massif—which straddles northeastern Greece and southern Bulgaria. The warm highlands of Albania and northern Greece are among the few European localities favored by the crested porcupine. Farther south the mountains of Greece are dry and thinly forested with conifers, broad-leaved evergreens and

maquis scrub. Devastated for centuries by hungry goats and an unforgiving climate, they are relatively poor in wildlife.

The Apennine Mountains of Italy rise generally to heights of 3000–5000 feet in the north, rather higher in the middle and toward the south. Composed mostly of Mesozoic and early Tertiary folded rocks, they have a partial overlay of marine deposits acquired while the area was submerged in late Tertiary times, and have risen relatively recently to face harsh erosion by wind, rain, and ice. Their lower slopes are forested with beech, chestnut, and oak, higher slopes with fir and pine opening to mountain pasture; though bitterly cold in winter, they carry no permanent snow or ice. The fauna of the Apennines is similar to that of the Alps but impoverished by intense hunting pressures. In the Abruzzi National Park of central Italy there are local races of the brown bear and chamois, and remnant populations of other mountain mammals. Birds are more plentiful, including black redstarts, redbacked shrikes, blue rock thrushes, alpine accentors, choughs, and crag martins.

The Pyrenees, which cross the neck of the Iberian Peninsula, are an ancient and complex range of mountains, largely worn down before being once more uplifted by the earth movements that raised the Alps. At the wetter, Atlantic, end they are heavily forested with chestnut, oak, and beech, and capped with pine woods and rich subalpine pasture. The drier eastern end is relatively bare, supporting only maquis and open woodlands of evergreen oak at lower levels, but conifers and meadows higher up. The highest peaks, in the central massif, rise above 10,000 feet and are snow-covered throughout the year. Southern-facing Spanish slopes are warm in summer and ice-free; north-facing French slopes are cooler and carry a few small remnant glaciers among the high peaks. West and south of the Pyrenees are the many lower but still impressive sierras of Spain and Portugal, mostly of prealpine origin except for the Sierra Nevada in the far south. The southern mountains are generally drier and more sparsely forested than the western Pyrenees, and have lost much of their wildlife. In the Pyrenees themselves you can still see a splendid selection of European mountain mammals, including such rarities as the Pyrenean desman (a pale, aquatic molelike creature inhabiting mountain streams), Miller's water shrew, the alpine shrew, pardel lynx, and Spanish ibex. These mountains are also remarkable for their populations of raptorial birds; especially notable are the lammergeiers, griffon and Egyptian vultures, black and red kites, Bonelli's eagle, booted and short-toed eagles, and several species of harrier, falcon, and owl.

The mountains of Europe's far north have a special character of their own. Here, rugged uplands, indented by deep, steep-sided, sea-flooded valleys called *fjords*, dominate the Scandinavian peninsula. Its western flank is a complex massif of old, hard crystalline rocks. Caught up in the Caledonian orogeny of 400 million years ago, and later lifted, faulted, and eroded by weather and ice, the massif now forms a long, high plateau, carved and dissected by steep-sided valleys. The eastern flank of the peninsula, made up of Paleozoic and younger sediments on a platform of older rocks, escaped the Caledonian and later folding but suffered

Reindeer (close kin of the American caribou) resting in Norway's far north. Most live in herds tamed by Lapps, but wild reindeer—preyed upon by wolves—roam Norway's mountains.

similar uplift, faulting, and erosion. Its contours are gentler, with broader coastal lowlands; its many upland lakes, moraines, and valleys with U-shaped cross sections are reminders of the ice-sheets that spread repeatedly over the whole peninsula during the past million years. The highest Scandinavian peaks occur in the Jotunheim region of southern Norway, where several exceed 8000 feet; the Norrbotten area of northern Sweden has peaks approaching 7000 feet.

Northern Scandinavia is crossed by the Arctic Circle, but the warm Norwegian Current saves the peninsula from polar conditions. Although snow lies for months in winter, year-round snow occurs only above 5000 feet in the south and above 3300 feet in the north; only in the far north do glaciers from the small inland ice-sheets reach down to the sea. The characteristic vegetation of Scandinavia is coniferous in which Scots pine and spruce predominate. On the lower hills of the south there are mixed deciduous forests of oak, beech, lime, ash, and birch, which merge into the pine and spruce forest on the higher ground. Upland birches and alpine grassland occupy the higher ground, giving way to tundra immediately below the zone of permanent snow. The limit of sizable trees occurs at about 2300–2600 feet in the south, falling to 650 feet in the far north where dwarf birch and willow, heather moorland, and lichens take over.

Although bleak and inhospitable in winter, the Scandinavian mountains have much to offer mammals and birds during the long days of summer. The mountain forests have been cut for timber or cleared for grazing, but wildlife is still abundant in the remoter regions.

Beginning in May, snow melts and the open meadows and tundra burst into a riot of color. Buttercups, cinquefoils, stitchworts, rock-cresses, vetches, cranesbills, and many other flowers bloom briefly up to midsummer, to be replaced almost immediately by red and golden autumn tints. The small herbivorous mammals include blue hares, Norway lemmings, gray-sided, short-tailed, and ground voles. Larger grazers include roe deer, occasional red deer in the coastal mountains of the south, and moose (called elk in Europe), which favor marshy woodland in summer and higher forested regions in winter. Reindeer, both wild and domesticated, are

Cliffs pockmarked by Buddhist caves and rock-cut figures line the poplar-studded Bamian River valley in Afghanistan. Behind rises the Hindu Kush, one of Asia's highest ranges.

plentiful in the far north and also in the southern mountains of Norway. Musk-oxen have been introduced into the Dovrefjell mountains of Norway, where they appear to have settled contentedly. Despite hunting pressures there are still wolves and bears to be seen in several of the national parks, and otters, beavers, wolverines, lynx, and martens are locally abundant. A striking feature is the summer abundance of migratory birds that winter in temperate Europe and Africa and fly north to the mountain pastures, marshes, lakes, and tundra in spring. They include many songbirds familiar both as passage migrants and as winter residents in Britain, and numerous water birds—ducks, geese, divers, and waders including dotterel, whimbrel, redshank, dunlin, and stint—that breed in the high moorlands, tundra, and lakes of Scandinavia.

Our brief survey has concentrated on the higher mountains within Europe. Mountain life is less fully developed in the lower highlands—for instance those of the British Isles, where no peaks much exceed 4000 feet. Then, too, man has here destroyed native mountain forests of oak, Scots pine, and birch. In their place stand boggy moorland and plantations of imported species of conifer. However, the high Scottish hills support a mountain grassland. Scottish uplands, too, harbor a distinctive collection of mammals and birds including red deer, blue hares, golden eagles, red grouse, ptarmigan, and dotterel.

We now turn from Europe to Asia, the great bulk of the Eurasian landmass, containing some of the mightiest mountains on earth.

The northern heartland of Asia is a stable platform in the USSR, bounded in the west by the Ural Mountains, in the east by the Verkhoyansk Range, and in the south by an arc of high ranges—the Altai, Sayan, Yablonovy, and Stanovoi mountains—which separate Siberia in the USSR from Mongolia and from Manchuria in northeast China. South of these mountains arid tablelands occupy much of Mongolia and of Sinkiang in northwest China. These tablelands are bounded in the west by the Tien Shan and Pamirs and in the south by the Karakoram Range, the Kunlun Shan and ranges projecting from the Kunlun Shan eastward through China. The lofty plateaus and ranges of northern and central Asia were formed originally by folding during the Caledonian and Hercynian orogenies of Paleozoic times, and have repeatedly been squeezed, faulted, and lifted during later periods.

Northwest of Mongolia the Altai and Sayan mountains respectively reach 15,000 and 11,500 feet above sea level. This is a region of continental steppe and mountain forest, with hot summers and intensely cold dry winters. Birch, poplar, aspen, and Siberian larch clothe the mountains to a height of about 8000 feet, and the alpine meadows above support gentians, lilies, primulas, columbines, and many other flowering plants with European affinities. The higher ranges south of Sinkiang rise from a tableland of 6500 feet to form a broken, towering wall over 16,000 feet high. Several peaks of the Kunlun Shan exceed 20,000 feet and some stand above 23,000 feet; the Karakoram Range is even higher, with peaks of 26,000 feet and more.

South of the Kunlun Shan lies the high plateau of Tibet, a faulted and dissected upland averaging over 16,000 feet in elevation. Despite its subtropical latitude, Tibet stands close to the level of permanent snow; it was heavily glaciated during the Ice Age, and its 20,000-foot peaks often carry a dusting of snow. It is a dry, semidesert area receiving only about eight inches of precipitation (in rain and snow) per year, and losing much of its moisture in evaporation under the fierce sun. Thin alpine grassland and shrubs form its main cover, with richer meadow and forest in the damper areas among the high valleys. Browsing and grazing animals include small flocks of chiru or Tibetan "antelope," with goas (smaller, gazellelike browsers), musk deer, wild blue sheep or bharals, wild asses, and heavy-coated yaks that roam widely both wild and in domesticated herds. Takins (heavily built goat-like creatures with swept-back horns recalling those of musk-oxen) browse in the wooded areas, and wild goats and argali (Marco Polo sheep) occupy the higher and poorer ground. Marmots and pikas also occur in their customary habitat among the crags and high scree slopes. The splendidly coated snow leopard and smaller Pallas's cat are feline predators. There is also a small weasel, which feeds mainly on birds and carrion, and Tibetan blue bears are occasionally seen wandering in their perpetual search for food.

Far and away the world's highest mountains, the Himalaya forms a stupendous southern wall to the Tibetan Plateau and an even more magnificent backdrop to the plains of India and

Above: blood pheasants in snow, 14,000 feet up in Nepal. These and other pheasant species are among the most colorful birds of the high forests in the Himalaya and nearby Asian ranges.

Pakistan. Over 1500 miles long and reaching heights up to 29,000 feet, the Himalaya rises from the plains in a series of three giant steps. The southernmost ranges, forming the Outer Himalaya, are forest-covered hills averaging a little over 3000 feet high. The second step—the Lesser Himalaya—is a complex system of folded and overthrust mountains with forested base and snow-capped peaks, rising to 16,000 feet and more. The main range is the Greater Himalaya, averaging 20,000–23,000 feet above sea level. Heavily capped with ice, it includes vast areas of alpine tundra and snow desert. Most of the major peaks beloved of climbers—Nanga Parbat, Kolahoi, Nanda Devi, Annapurna, Everest, Makalu, Kanchenjunga and Namcha Barwa—form part of this tremendous range. The Greater and Lesser Himalaya are made up of ancient crystalline and metamorphic rocks, overlain by marine sediments from the Tethys Sea that were injected with molten rock, covered by submarine eruptions of lava, and folded under pressure during the early and middle part of the Tertiary period. The southern ranges are younger, composed mostly of coarse gravels and sandstones derived by weathering from the older rocks to the north, and subsequently consolidated and folded by later pressures.

The Himalayan ranges are almost encircled by two great rivers—the Indus and Brahmaputra—which rise close to each other among the glaciers and snowfields on the northern faces of the

Climber dwarfed by compacted snow and ice in the Western Cwm, a great ice-worn rock below Mount Everest. Few mountain organisms are adapted for prolonged survival in such places.

Great Himalaya. Hundreds of small rivers cut the southern faces of the mountains, and those of the central ranges are caught by the Ganges, which flows southeastward along the southern foot of the mountains to the Bay of Bengal.

The Himalaya forms a northern boundary to the monsoon area of India and southeastern Asia. In summer, when the sun is overhead and atmospheric pressures are low over the Asian mainland, moist southerly and southwesterly monsoon winds blow in from the Indian Ocean, bringing heavy, persistent rain to the Himalayan foothills. The west is generally drier than the east, and the inner ranges are both drier and cooler than the southernmost, so there are many local variations of vegetation and animal life

Below: langur monkey, one of the many animal species once numerous in the southern foothills of the Himalaya, but now being driven out as men fell forests and use the land for farming.

over the Himalaya as a whole. In the warm, damp monsoon climate the southern hills grow dense tropical or monsoon forest. This formerly rose in a continuous sweep from the plains, and was the home of the Indian elephant, Indian rhinoceros, langur monkey, wild pig, and tiger, leopard, jackal, mongoose, and civet. The forests of the foothills were especially rich in different species of deer, including barasingha or swamp deer, chital or axis deer, hog deer, sambar, and muntjac, which generally inhabited the wetter areas where year-round food was abundant. Much of the low hill forest has now been destroyed, and the ground from which it sprang has been terraced laboriously for rice cultivation. Animals still roam wherever the forest remains, though hunting pressures have reduced many local populations to danger levels and beyond.

From about 8000 feet upward the forest changes, opening to cooler and drier woodlands of deciduous or evergreen oaks, chestnuts, and tree rhododendrons, with shrub rhododendrons, azaleas, and bamboo forming a dense subcanopy. Grasslands and coniferous or mixed woodlands replace the deciduous forest in drier areas, producing an attractive, open parkland with a rich variety of colorful shrubs and herbs. In wetter regions at cloud level, ferns, epiphytic mosses, and lichens drape the trees. Pines, cedars, and firs become more prominent on higher slopes, with dense stands of bamboo in the wetter areas and tall cypresses and other conifers on drier

Lammergeier soaring 12,000 feet up in Nepal's Greater Himalaya mountains. Such denizens of wild, inaccessible heights will outlast the more vulnerable creatures of the foothills.

ground. Higher still small birches, rhododendrons, firs, and other trees predominate up to about 13,000 feet. Above lies a treeless zone of dwarf rhododendrons, dwarf willows, birch shrubs, and juniper. On southern slopes this zone of dense but restricted growth extends upward to meet a permanent snow line at about 14,800 feet. The highest peaks which are generally drier and subject to bitter winds from Mongolia, have a higher snow line at about 20,000 feet with extensive areas of snow-free rock faces extending up to 23,000 feet and above. Here a broad band of alpine tundra intervenes between the zone of shrubs and the permanent snow, with a flora of sedges, coarse grasses, primulas, gentians, edelweiss, and other flowering plants, many of which grow in characteristic cushion formation.

Lawrence Swan, an American biologist making a special study of high mountain environments, has described Himalayan conditions at 15,000 feet and above as like a prolonged mild winter; there is a long season in which growth is possible, but growth is restricted in the wetter regions by permanent snow cover and in the drier regions by constant shortage of water. Plant cover is patchy but varied, and provides shelter and food for many small animals including insects and spiders. Butterflies, bumblebees, and craneflies—often dark-bodied to absorb heat—fly busily when the sun shines but rapidly become torpid in the shade; in cloudy conditions, and in the cool of early evening, they fall prey to wolf spiders, whose insulated "hairy" bodies seem less sensitive to temperature change and able to continue working efficiently at lower temperatures. Swan found bees, wasps, ants, flies, beetles, bugs, aphids, stoneflies, grasshoppers and thrips (small sucking insects) common up to about 16,000 feet. Several observers have commented on the prominence of moths and butterflies in the high Himalaya. Butterflies include a dozen Apollo species (many of which occur at 15,000 feet and above) and several kinds of blues, ringlets, vanessas, papilios, and whites, generally at 10,000 to 15,000 feet in the forests and grasslands.

Above 18,000 feet is a truly arid zone with few plants of any kind. Although snow may fall at any time of the year it is never plentiful and much of it vaporizes directly into the thin mountain air; only a small fraction melts to dampen the ground, and green plants grow only in crannies where moisture accumulates. The highest green plants of the Himalaya, and probably of the world, grow in such situations a little above 19,700 feet. Collembolans (springtails) and thysanurans (a group of primitive insects that include the household silverfish) occur in crevices and under stones up to over 16,000 feet, even managing to survive in the barest parts of this mountain desert region. Swan recorded thysanuran insects living actively on warm rock at 19,000 feet. Mites and small centipedes were found at 19,350 feet and jumping spiders were recorded by one of the earlier Everest expeditions at 20,000 feet. How animals live so far above the limit of vegetation is not entirely clear, but we know that updrafts of air sweep pollen, seeds, dead and dying insects, and other organic debris up to this level. Here these materials form the basis of food chains for tiny scavenging insects, mites and spiders.

Small mammals and birds are plentiful in the higher forest zones of the Himalaya, and a few species extend beyond to the limit of vegetation. Within the bamboo thickets are bamboo rats, which burrow and feed mainly on the roots and young shoots of the bamboos. Other small rodents living above 13,000 feet include mice, Père David's voles, bobak marmots, and flying squirrels. Pikas have been recorded almost to 20,000 feet, but they normally live where vegetation is more plentiful. Snow partridges and Himalayan choughs breed above the timberline, and snow cocks, Hume's ground jays, desert chats, ravens, and other ground-feeding birds scratch a living in the alpine zone. Griffon vultures, lammergeiers and golden eagles soar in the up-currents, scanning the mountain slopes for small mammals and carrion. Tibetan weasels are the main predators of ground-living birds, taking their eggs and nestlings; foxes and Pallas's cats hunt the small mammals, and scavenge where larger mammals have fed.

There are many species of larger grazing and browsing mammals, some restricted to the high forest, but most wandering through forest and alpine zones, taking food wherever they find it. Largest and most striking are the yaks—sturdy black cattle with short limbs, long horns, and long, fine coats. In summer they graze at 18,000 feet and above, feeding on long grass and shrubs. Yaks have long been domesticated, but many still live wild in small wandering herds. Blue sheep (bharals) and the smaller urials feed on mountain pastures at lower levels, and gorals—small, stocky goat-antelopes—range widely on rocky hillsides where the grass grows well. The hangul or Kashmir deer—a local variety of red deer—and the very much smaller musk deer feed in the forest, emerging on the higher pastures in summer. The high reaches of the forest support short, triangular-horned goats called tahr, which graze in bands on the steep, tree-clad slopes. Above them, on the exposed crags of the alpine zone, live markhor—larger and more splendid mountain goats with shaggy brown hair and corkscrew horns—and ibex, which, as always, inhabit the highest slopes of all in the regions of permanent snow. Wolves and snow leopards are the main predators of these herbivores, though eagles, lammergeiers, and other raptorial birds probably account for many of their young. Black bears, mainly confined to the forests, may also attack and kill some of the forest-dwelling herbivores, though like bears the world over they also feed lazily on roots, insects, fish, honey, eggs, and carrion. Brown bears play a similar role of scavenger on the higher slopes.

At the eastern end of the Himalaya, and forming the eastern edge of the great plateau of Tibet, stand the rugged old mountains of China's Szechwan Province. Formed of ancient limestones and red sandstones, they resisted crumpling during the Himalayan orogeny, but were lifted and subsequently carved by glaciers and streams. Now they are deeply dissected by fast-flowing rivers including four great streams that unite to form China's Yangtze River and the Salween and Mekong rivers that flow south respectively into Burma and Thailand. Heavily forested with bamboo, rhododendron, and conifers up to 16,000 feet, these mountains are the home of several rare or unusual animals. Best known in the popular sense is the giant panda, whose engaging black-and-white face and chubby contours have become the symbol of all endangered wildlife. Giant pandas live in the bamboo forests of Szechwan at elevations of 4500–9000 feet. Their front paws, with a curious pad like an extra thumb on the inner edge, are specialized for grasping bamboo stems; large teeth and powerful jaws cut and crush the stems with remarkable efficiency. Although restricted to a narrow ecological zone on a few neighboring mountain ranges, giant pandas are probably protected by their remoteness from civilization, and may not be in immediate danger of extinction. Red pandas, smaller and more raccoonlike, are native both to the mountains of western China and to the Himalaya. They range fairly widely in the mountains, feeding on fruits, seeds, bamboo shoots, and roots, which they dig up with sharp claws and teeth.

The deep valleys of Szechwan and neighboring areas also hide many species of browsing mammals, including several distinctive but related forms of goat antelope. Takins inhabit rhododendron and bamboo thickets on the higher slopes in summer and descend to the valleys in winter. Szechwan takins, which are yellow in summer and gray in winter, were once widespread but hunting has much reduced their numbers. Golden takins, restricted to high forested crags in the Tsinling Mountains of southern Shensi Province, seem also to have reached a dangerously low level of population and are now protected. Gorals, which we have already met in the Himalaya, graze on high mountain pastures close to the timberline. Surefooted and swift, they are usually found near crags and rough rock that assure them a ready means of escape from predators. Serows, black-coated and goatlike, are widely distributed in deep woodlands of China, especially in steep-sided valleys which few other large browsing animals can penetrate. Père David's deer, known only in captivity when discovered by science, may once have inhabited the wooded lower slopes of these mountains. The Shansi sika, a large, faintly-spotted deer that once inhabited the higher mountain forests northwest of the Chinese heartland, is almost certainly extinct.

Lack of space prevents us from exploring all Asian ranges. But this chapter has at least sketched mountain life in most of the representative mountain masses of Eurasia.

West of the Himalaya the knotted ranges of fold mountains sweep through Afghanistan and Pakistan, skirting the Arabian Sea, and continuing north and west through Iran and Turkey to link with the Caucasus and the mountains of southern Europe. Afghanistan's southern plateaus are dry and dusty deserts, but the high Hindu Kush in the north carries walnut, evergreen oaks, and various Asian conifers up to 13,000 feet. Animals here are like those of the Himalaya. Farther west, the Zagros Mountains of southern Iran and the cloud-making Elburz Mountains of the north are forested to 6500 feet with trees more familiar to western eyes—elm, maple, hornbeam, beech, lime, and ash. The Caucasus Mountains, forested to a similar level, stand at the boundary of east and west; their trees are largely western but animals represent a strange geographical mixture. There are bison, ground squirrels (susliks), jerboas, jackals, a very few leopards, wild boar, and long-clawed mole-voles native to the area. Commonplace mammals with European affinities include bear, lynx, fox, red and roe deer, chamois, and ibex.

Female giant panda. These bear-like mammals related to raccoons live only in the mountain forests of Szechwan in south-central China and in nearby Tibet. They chew bamboo leaves and stems, herbs, and they also eat small rodents.

Africa

Unlike most continents Africa shows evidence of a long history of geological peace and stability over much of its surface. It is basically a great slab of continental crust that has missed colliding with other continents except in its northwestern corner. So a great deal of Africa has suffered relatively little folding or down-warping, and has few sedimentary rocks of recent origin; the ancient crust forming much of the surface includes some of the world's oldest recorded rocks.

Africa's average elevation is over 2000 feet, and the land is mainly high plateau. There is only one major range of fold mountains—the Atlas Mountains of Morocco and Algeria. Although lacking the vast transcontinental chains of the Americas, Asia, and Antarctica, Africa has spectacular mountains of other kinds. The high upthrust massif of western Ethiopia, the smaller block mountains and volcanoes of East Africa, and the dissected peaks and tablelands of the south provide a wide selection of mountain scenery. Largely because of their isolation from each other, Africa's mountains have acquired a strange variety of plants and animals—some curiously specialized for alpine life on or near the equator.

Through its early geological history Africa formed the core of the Gondwanaland supercontinent. For many millions of years it occupied a polar position, supporting a huge ice-cap, extensive glaciers, and tundra vegetation, all in regions now occupied by hot desert and tropical rain forest. As Gondwanaland drifted away from the South Pole, Africa and its neighboring lands warmed, becoming tropical about the time the supercontinent began to break up in the middle and late part of the Mesozoic era (the era lasting from about 230 to 66 million years ago). The dispersal of Gondwanaland carried South America westward and Antarctica and Australia eastward, leaving Africa relatively intact and more or less in its present shape. For millions of

Suffused sunlight deceptively softens the outlines of a rocky scene in northwest Africa's Atlas Mountains, the continent's only major group of young fold mountains. Berber farmers have tamed this valley floor, but wildlife holds the slopes above.

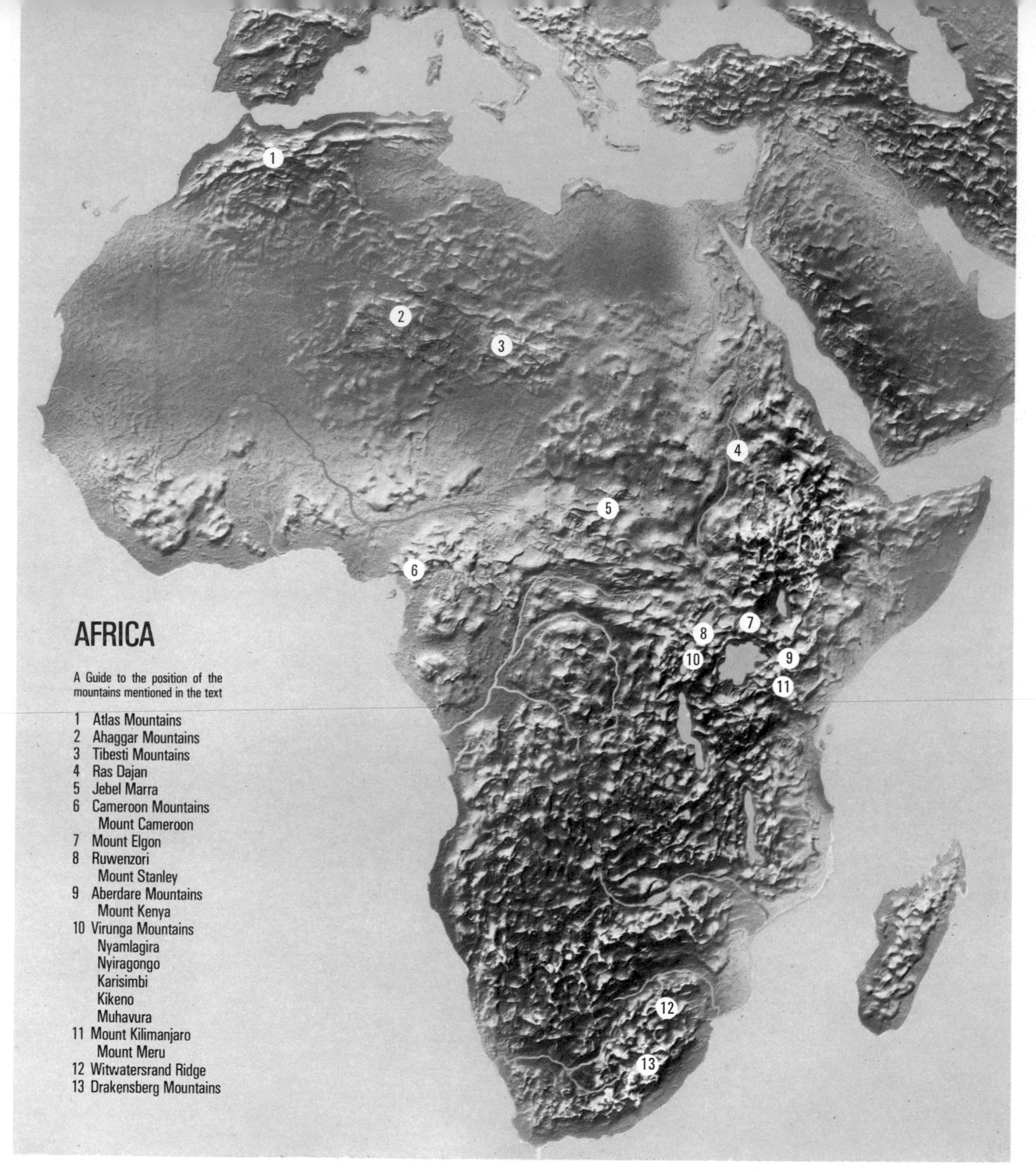

years in Mesozoic times northern Africa was probably heavily forested—part of a forest system that extended with breaks to the eastern extremity of the continent. As a result of this Afro-Asian heritage, many African insects, birds, and mammals (including chevrotains or "mouse-deer," scaly anteaters, porcupines, and some monkeys) show close kinship with Oriental species. But during the Tertiary period (from 65 to 2 million years ago) climatic changes replaced vast tracts of northern African forest with savanna and desert. This tended to isolate the central and southern forests from those of Asia, so that Africa south of the Sahara developed its own distinctive forest flora and fauna from which modern mountain species evolved.

During middle and later Tertiary times pressures developing between adjacent African and Eurasian continental plates gradually thrust up the Atlas Mountains, simultaneously with the Alps and other fold mountains of Europe. Most other African mountains are also relatively

Apart from the Atlas ranges, most African highlands are largely of volcanic origin, for instance the Ahaggar and Tibesti ranges, Mount Cameroon, and Great Rift Valley mountains (8–13).

new. Some, like the Ahaggar Mountains of southern Algeria, the Tibestis of Chad and the high peaks of Cameroon, arose independently by volcanic upheaval. But many of the eastern mountains are linked with the Great Rift Valley system, which in its late stages of development produced vast outpourings of lava and ash, and gave rise to spectacular volcanic cones in Kenya, Uganda, and Tanzania.

For our first look at African mountain life we shall examine the northern ranges and blocks, starting with the majestic Atlas Mountains that stand between the Mediterranean Sea and the Sahara Desert. These are fold mountains that probably reached their greatest height some 30 million years ago and have since been partly worn down. The folds run approximately east-northeast to west-southwest, and their line is continued east-northeast through Sicily and southern Italy and west-southwest through the high peaks of the Canary Islands. During the Ice Age the Atlas Mountains were heavily glaciated. Today many of their highest peaks above 11,000 feet carry snow for much of the year, but little permanent ice remains.

The highest mountains of the Atlas ranges stand in the High Atlas region, south of Marrakesh in Morocco. Here, several peaks exceed 11,500 feet, and some rise almost to 14,000 feet. The rocks are hard, well-worn sandstones, limestones, and schists, with marble, diorite, basalt, porphyry, and a core of granite revealed on some of the southern exposures. South and east of the High Atlas, ranges of lower mountains linked by plateaus reach elevations of 3000–8000 feet. The Maritime Atlas flank the Mediterranean coast of Morocco and Algeria. The Saharan Atlas of the south merges into the high, rolling plateau lands of the Sahara Desert. The western Atlas Mountains have a Mediterranean climate, with reliable winter rainfall coming in from the Atlantic Ocean. Their seaward slopes are forested with evergreen and cork oaks, Atlantic cedars, and Aleppo pines, and the high tablelands between the ridges have for many generations been cleared and cultivated by peasant farmers. Alpine tundra caps many of the highest peaks. Slopes of the Saharan Atlas tend to be drier. Desiccated by hot winds from the Sahara, they carry only an impoverished steppe or desert vegetation.

North Africa and Europe have been joined repeatedly in geological history, which explains why the Atlas Mountains have European as well as African animals. In both cases big mammals have suffered from hunting and destruction of habitat. One species of an Algerian bear and another of wild ass disappeared some time ago, and the last Barbary lion was killed by hunters during the 1920s. The Barbary deer, a local variant of the European red deer, survives in small local populations high in the meadows and forests of the Maritime Atlas. Heavy-coated Barbary or aoudad sheep, with goatlike horns, graze on the high meadows and crags, and delicate dorcas gazelles range in small bands over grassland. Wild pigs are still fairly plentiful in the mountain forests. Leopards and hyenas, both widespread in northern Africa, are the remaining large carnivores of the uplands.

Several smaller species of carnivore inhabit the Atlas, feeding mostly on small rodents, birds, and reptiles. Because their food is scarce and widely dispersed, they tend to be nowhere numerous. The fennec, a desert fox with enormous ears, occurs throughout the northern deserts of Africa and Arabia, and has adapted well to mountain conditions. Jerboas and gerbils—small, nocturnal desert rodents—are probably its main prey, though fennecs also eat insects and small reptiles. European wildcats, elegant tawny caracals (desert lynx), and genets—small catlike carnivores related to the mongoose—also prowl at night in bush-covered regions of the Atlas Mountains, and golden jackals hunt in small groups in wooded and open country.

Barbary apes (tailless monkeys, not true apes) are plentiful in the high rocky forests of the Atlas. Best known from the small, pampered population on the Rock of Gibraltar, they live in small, scattered bands among crags and trees, hunting by day and sleeping in sheltered corners away from the cold night air. They feed omnivorously, eating leaves, young shoots, berries, pine seeds, fruit, insects, caterpillars, and even small mammals. Constantly watchful, they are notorious raiders of crops and food stores, and heartily disliked by the hard-working peasant farmers of North Africa. Their closest kin are the macaques of India and Southeast Asia.

Other characteristic creatures of the Atlas

include eagles, vultures, and smaller birds of prey: hawks, falcons, and owls.

The Ahagger and Tibesti mountains lie in the central Sahara Desert respectively south and southeast of the Atlas Mountains. These are volcanic peaks that rise to 10,000 feet and more from dry, desert platforms of Paleozoic rock. Because of their altitude they are cooler and damper than the surrounding hot desert, and carry a thin cover of drought-adapted plants including grasses and scrub. They are thus more hospitable than the nearby lowlands, and tend to attract animals. The largest creatures that these *hamada* or high desert areas support are the addax, a heavily built antelope with long spiraling horns; the scimitar oryx, a large antelope with slender, curved horns; and the dorcas gazelle. There are also flocks of grazing ostriches, and small, wandering herds of wild asses make a meager living from the poor vegetation. There are no large predators on the *hamadas*. Most of the wildlife is small enough to live under the scant vegetation and dig into the sand to escape the burning sun. Jerboas (jumping mice with long hind legs) and fat sand rats (a species of gerbil that stores fat con-

Right: long, sensitive ears assist the dorcas gazelle to detect approaching danger. Numbers of these dainty, soft-eyed, swift-footed antelope graze upon grassy regions in the Atlas Mountains.

Left: the Moroccan mountain town of Boumalne, just south of the High Atlas. Melting snow from distant peaks feeds the Dadès River, bringing fertility to a valley besieged by harsh, dry hills.

spicuously under its skin when food is plentiful) feed on roots and young vegetation. Desert hedgehogs are insectivorous, taking locusts and other herbivorous insects. Skinks (smooth-bodied lizards), horned vipers, and puff adders are among the carnivorous reptiles that live on the *hamadas*, and their birds include spotted sandgrouse, desert and bifasciated larks, and lanner falcons.

Jebel Marra, a mountain massif rising above 10,000 feet in western Sudan, rises from dry acacia woodland and savanna. With a rainfall of about 30 inches per year it is considerably damper and more habitable than the desert mountains to the north. Deep stream-cut valleys testify to an even wetter climate in the recent past. The open slopes are mostly clothed in grasses and low shrubs. Valleys contain rich accumulations of soil, and wild olives, figs, and acacias grow here, with willows and clematis on the stream banks.

The soil and luxuriant vegetation support a surprisingly rich fauna of wood lice, insects, spiders, and mites. Over 35 species of mammals—some unknown in the surrounding lowlands—have been recorded from the massif. These include two species of bats, two tiny shrews, leopard, cheetah, giraffe, African elephant, African lion, and over a dozen species of rodents. Local forms of gerbils and striped mice have been found at the highest levels living among bracken near two crater lakes. The lakes themselves contain sizable populations of dabchicks. Greater kudu, duikers, and red-fronted gazelles graze the lower slopes where hyenas and hunting dogs prey on them. Smaller carnivores of the *jebel* include bush cats (wild African tabbies probably ancestral to domestic cats) and zorillas—striped polecats that are related to weasels and feed mainly on small rodents. Small colonies of rock hyraxes occur in the high grasslands above 3000 feet. These rabbit-sized grazing mammals have hoofed toes, and live in burrows and rock cavities. They are sociable little animals that feed, drink, travel, and sun themselves in groups, keeping to well worn tracks and grazing in fan formation, as though constantly on the alert against predators; one individual's high-pitched alarm call will send a whole group running for shelter. Rock hyraxes feed rapidly, mostly on the leaves and young shoots of woody plants which they cut with sharp side teeth, but also on grasses and softer material. Wandering colonies of dog-faced baboons, green guenon monkeys, and patas monkeys have also been seen on the Jebel Marra, generally on the grasslands and rocky outcrops of the lower slopes.

Africa's most dramatic mountain scenery occurs some 1000 miles to the east, in the high tablelands of Ethiopia. Here an enormous block of ancient metamorphic rock, capped with limestone, sandstones, and volcanic basalts and dolomites, has been lifted bodily to stand high above the general level of the African plateau. Deeply cut by ravines, the block is itself topped by a broken plateau between 6000 and 10,000 feet high, with many snow-covered peaks reaching heights of 13,500 feet and more. The highest peak, Ras Dajan, stands over 15,000 feet tall and is permanently capped with snow. The canyons and deep gorges, some with sheer walls over 3000 feet high, and the remote crags and rainwashed peaks make Ethiopia a difficult country for colonization, whether by plants, animals, or men. Heavily eroded by ice and denuded of life during the glacial period, its mountains seem still to be underpopulated and awaiting the arrival of new species from outside.

Rock hyraxes warily face outward on a rocky knoll rising like an island from a sea of trees. These small, hoofed, mountaineering mammals live in Africa east and south of the Sahara.

The high valleys of the massif tend to be warm, damp, and heavily forested with a wide range of subtropical trees and shrubs. The more accessible parts of the forest have been destroyed for their timber. But wild olives, giant sycamores, wild pines, cedars, and gums, grow in profusion, with date palms and bananas, citrus fruits, pomegranates, and coffee flourishing both wild and under cultivation in the fertile black soil. Above 11,500 feet the forest thins to grassland and heath, dominated by giant heathers and lobelias. The tall heathland vegetation often carries a burden of moss and other epiphytes that grow well in the damp air. There are extensive plateaus of grasses and herbs, often cleared for farmland and suffering erosion from over-intensive cultivation. These fertile uplands are readily reached from below by birds. Several species, including white-collared pigeons and red-winged starlings, fly up each morning from the lowlands—a vertical distance over 3000 feet—to feed on the abundant seeds or insects, returning each evening to the warmth of the valleys below. Large birds of prey take advantage of morning updrafts to feed on the heights; tawny eagles, lammergeiers, and Egyptian vultures nest on inaccessible ledges of the cliffs, rising into the morning sunlight to scan the plateaus for live prey and carrion. Guinea fowl, partridges, sandgrouse, and other ground birds scratch a living on the high grasslands. Thick-billed ravens, like their crow kin the world over, are general scavengers that live by their wits on a wide variety of insects, berries, and carrion. Double-collared sunbirds are perhaps the brightest birds of the uplands; iridescent green, blue, and yellow, they flit like some of the larger species of hummingbirds from flower to flower in search of nectar and insects. These are among the few species of birds that allow their body temperature to fall during the night, presumably as a means of conserving energy.

The most prominent inhabitants of the highlands above forest level are the geladas, which live on cliffs and crags of the heath and grassland areas. Each night these baboons or baboon-like monkeys (experts cannot be sure which they are) retire to roost in small family parties, usually of one male accompanied by females and young. Thick brown fur, with a distinctive mantle over the shoulders, protects them from rain and cold. At first light they move up onto flat open ground, and spend the day there feeding and playing. They seem to eat almost any handy living material including leaves, stems, roots, tubers, seeds, fruit, birds, small mammals, lizards, eggs, insects—even scorpions. The troops move sociably, their members communicating with calls and grunts. Sometimes they combine forces to catch hares or small rodents. Always

Farms occupy much of the Ethiopian plateau, as here, 8000 feet up near Addis Ababa. But Ethiopia's uncultivated ravines and high heaths hold Africa's finest array of mountain life.

there are several pairs of alert, intelligent eyes on watch for predators. Their main enemies are leopards and big birds of prey, though jackals, hyenas, and other large carnivores often cause the troops to disperse and race for safety. Each young gelada rides under its mother, clinging tightly to her fur for the first few weeks after birth. Later it rides upright on her hindquarters like a jockey, holding tightly with knees, feet, and hands while she negotiates the steep cliffs and crags of their mountain home. Doguera and hamadryad baboons are also found in the Ethiopian bush country, though seldom as high as the geladas.

Klipspringers (small antelopes with a rough, yellow-brindled coat) and mountain nyalas (heavier antelopes with long, gray-brown fur and a dark mane) graze and browse on the heathlands in small herds. The sure-footed klipspringers favor rough, rocky country, where they practice their skills of leaping among pinnacles and ledges, whereas nyalas live more soberly in the dense vegetation on flat ground. The highest crags are the home of the walia or Ethiopian ibex, a sturdy brown mountain goat with magnificent knurled horns. The walia was once widespread but is now restricted almost entirely to small areas of the Simen Mountains. Hyraxes and hares are plentiful on the grasslands; their predators include the black-tailed Simienen fox and other small carnivores. In the eastern highlands the Abyssinian genet, a catlike member of the civet family, is a vigorous predator of small birds, mammals, and insects. West of the rift valley that divides the Ethiopian Highlands the blotched and feline genets have a similar role.

South of Ethiopia, the highlands of Kenya and northern Tanzania owe their origins to the crustal movements and volcanic activities that formed the Great Rift Valley system—a system of rifts and cracks, produced by intermittent tensions in the earth's surface over millions of years.

The rift begins in the Jordan Valley of Israel and Jordan, continues southeast as the narrow, flooded channel of the Red Sea, then strikes south into the African continent, cutting a swathe southwest through the Ethiopian mountains. From here it divides into separate Eastern

and Western rifts that skirt the high plateau of Lake Victoria. The Eastern Rift cuts southward through Kenya and Tanzania, its route marked by many volcanoes, now mostly inactive. The volcanic mass of Mount Elgon (14,178 feet) lies to the west of the valley; to the east stand the Aberdare Mountains of west-central Kenya and the higher, more isolated and more dramatic peaks of Mount Kenya, Mount Meru, and Mount Kilimanjaro—Africa's highest peak at 19,340 feet. Steep escarpments line the valley. Though by no means continuous, they are often impressive. The highest land includes not only volcanic cones but immensely thick sheet lavas. In the uplands the atmosphere is damp, and the volcanic soil fertile. Thus the eastern mountains are well clothed with vegetation.

Rain falls seasonally in equatorial East Africa, following a monsoon pattern. The periods January-February and July-September tend to be dry, while March-May and October-December are wet. The plains and foothills are generally too dry to support forest, but grass grows well; this is the savanna of eastern Africa, famed for its massive herds of antelope and other game. Rain falls most heavily and persistently on the southern and western faces of the mountains, and year-round humidity is highest in a narrow zone 6500–10,000 feet above sea level. Typically, therefore, the eastern mountains rise from a dry plain but support a zone of montane forest starting at about 6500 feet. This zone is broadest and best developed on the wetter slopes but seldom extends far above 10,000 feet. During glacial periods the lowlands of central and eastern Africa were more heavily forested than at present, and the separate belts of forest that we now see on individual mountains were probably united. For this reason—and because of their basic similarity of structure and ecology—the various patches of mountain forest tend to contain similar wildlife.

Right: klipspringers (nicknamed "African chamois") are agile mountain antelope that occur from Ethiopia to South Africa They have rough coats and stand about 20 inches high at the shoulder.

Left: part of the crater rim of an extinct volcano near Kenya's Lake Naivasha, in the Eastern Rift Valley. A division of the Great Rift Valley, this crack in the earth's crust runs through a region of volcanic cones, and uplands largely formed from thick lava sheets.

The Aberdare Mountains are in many ways typical of the eastern highlands—a complex of lava sheets and long-dead volcanic cones rising to an average height of 11,000 feet from a platform roughly 5000 feet above sea level. The highest peaks at about 13,000 feet are periodically touched with snow. The Aberdares formerly supported dense forests of bamboo, cedar, camphor, juniper, yellowwood, also conifers of the podocarp family and many other sizable trees, with a dense understory of wild bananas, colorful shrubs, and bracken. Grassy clearings extended through the forest, blending into extensive moorland above 10,000 feet. Much of this vegetation has now been cleared for lumber and to provide land for cultivation, so that only patches of the original forest remain. You can still see forest mammals in reserves, though most kinds are quiet-moving and well-hidden, and many are nocturnal.

The large herbivorous mammals of the forest include elephant, rhinoceros, and buffalo, which are heavy enough to force their way through the tangle of undergrowth, and provide tracks for other mammals to follow. Giant forest hogs, too, are vigorous trailbreakers with small family parties barging through the dense vegetation of the forest floor in search of fruit and good grazing. There are several species of forest antelope, including the tiny suni, small red and larger black-fronted duiker, and spiral-horned forest bongo. Like many other mammals of the mountain forest, these antelope are smaller than their kin on the plains. They are mainly browsers and grazers, taking grasses, herbs, leaves, fruit, and young shoots from the forest floor and up to a height of six feet. Blue monkeys, colobus monkeys (with strikingly black and white fur), bushbabies and other primates make use of higher levels of the forest, browsing mainly on young shoots, leaves, and fruit. Leopards are the main predators of the larger mammals, hunting by climbing the trees and dropping on their prey from above. Crowned hawk eagles swoop dramatically through the treetops to strike at young antelope, snakes, and small mammals, which they lift out with powerful wings and grasping talons. The forest floor, litter-strewn all the year, provides cover and food for a variety of shrews and small rodents, the main prey of many of the small ground-living and tree-snakes, and of red-chested owlets, wood owls, and other predatory birds. Montane forest, often dense and gloomy despite the brilliant sunlight outside, is enlivened by colorful butterflies and many brilliant and richly hued birds. Fruit-eating touracos, hornbills and parrots, double-collared and other vivid species of sunbirds, bulbuls, and dozens of kinds of insectivorous birds—including flycatchers, forest robins, shrikes, robin chats, and thrushes and barbets—bring color and sound to every level of the forest.

Rosettes of giant groundsel on moorland high up on Mount Kenya. The shape and structure of these giant plants equips them to survive intense solar radiation by day, bitter cold by night—major hazards to life at high altitudes near the equator.

The higher zones of East African mountain vegetation show up most clearly on the shoulders of the huge, isolated volcanoes, Mount Kenya, Mount Kilimanjaro, and Mount Elgon, which all arose late in the Tertiary period and have now slumped into quiet though still spectacular middle age.

Mount Kenya (17,058 feet) rises from a base of dry, grass-covered plains, supporting rain forest from about 6500 to 10,000 feet. On the foothills and lower slopes, where annual rainfall seldom exceeds 40 inches, dark stands of cedar form the bulk of the forest. On higher ground these give way to dense thickets of hardwood, laced with lianas and bearded with lichens and mosses. In the upper forest zone of the southern and western sides, where clouds often gather and rainfall exceeds 65 inches per year, hardwoods give way to dense, impenetrable stands of bamboo. Despite their constant growth and hospitable appearance, bamboo thickets attract few animals in Africa; only elephant and buffalo seem weighty enough to break into them and, except in years of heavy seeding, they provide little food for small mammals and birds. The drier northern slopes of the mountain, where rainfall is generally below 35 inches per year, cannot support continuous forest. Here, the savanna and scrub that skirt the foot of the mountains continue with hardly a break to a height of 10,000 feet. At about this level Mount Kenya is ringed by a narrow zone of mixed forest and parkland, known to botanists as the *Hagenia-Hypericum* zone for the two species of broad-leaved trees that dominate it. Closely associated is a narrow belt of giant heaths and heathers, which grow to tree size in the damp atmosphere. On the western slopes these curious plants, suggesting a vastly overgrown moorland, are restricted to gullies and sheltered corners, but they form a broad band on open ground in the northeast and southeast. Giant hypericum (akin to Saint John's wort) and other shrubs form a patchy understory, with tussocks, rushes, and other marsh grasses covering the wet ground.

Above 11,500 feet Mount Kenya's upper slopes support a remarkable alpine meadow vegetation, dominated by plants found only at similar levels on other African mountains. From a continuous carpet of tussock grasses, alchemillas ("lady's mantle"), geraniums, mosses, and everlasting flowers grows a strange shrubbery—in places almost an open forest—of giant plants of the genera *Senecio* and *Lobelia*. These genera are better known for their small herbs, including the familiar groundsels, ragworts, and garden lobelias. On Mount Kenya's deeply dissected upper slopes this meadow and shrubbery alternates with the giant heaths and heathers. On the Aberdare and other mountains it forms a distinct zone of its own, covering wide expanses of high plateau and rolling upland to the tops of the lower mountains or, on higher peaks, to the snow line.

Some of the giant groundsels grow as squat, monolithic shrubs with rosettes of leaves half

a yard across and clusters of huge yellow flowers up to a yard high. Others form trees 15–18 feet tall, with bright yellow or purple flowers. The lobelias also take the shape of squat rosettes of stout waxy or hairy leaves with tall spikes of blue flowers rising from their center. The rosettes open during the day and close at night. When open, their shiny white surface protects them against excessive radiation from both the sky and the ground; when closed, their outer leaves protect the delicate growing tissues against night frosts. A mass of dead leaves about the base of each plant and thick corky stem give further protection, insulating the active layers of the plant against extreme heat and cold and making the whole plant less attractive to browsers and other animals. The open rosettes of some species trap water in the bases of their leaves, forming tiny pools in which midges breed. The hairy protective covering of many of these leaves is stripped away by malachite sunbirds, which use it as lining and insulation for their nests.

From about 14,000 feet the alpine meadow thins, only hardy green and golden mosses, lichens, and small flowering plants surviving in damp and sheltered patches among the rocks and semi-persistent snow. This high meadow resembles the high alpine zone of the European Alps. Many of its species of grass, saxifrage, rock cress, lily, and other small herbs are closely related to Alpine species and occupy similar situations. There are also identical and closely related species of insect, snail, and other invertebrates. However, the everlasting daisies, red-hot pokers, and other African plants point

Wildebeest on the plain below Africa's highest peak. Mount Kilimanjaro. Climbing Kilimanjaro, you pass through mountain forest, then moorland, then alpine plants, before reaching snow.

the difference, and the birds and small mammals are distinctively African—usually close kin of species that thrive on the plains below. This is probably a harsher environment than any found on the Alps; indeed many biologists regard the tops of equatorial mountains as among the most hostile climatic zones of all on account of the intense, relatively unfiltered radiation from the sun each day and biting cold, often with sharp ground frost, each night. Not surprisingly, Mount Kenya's highest crags are almost completely lifeless, with only sporadic lichens growing in favored spots among the snowfields.

Mount Kilimanjaro, three degrees south of the equator on the Kenya-Tanzania border, stands nearly 2300 feet higher than Mount Kenya. Kilimanjaro's upper slopes seem generally drier than those of Mount Kenya, and neither bamboo nor *Hagenia-Hypericum* zones occur above the montane forest. Instead there is a broad zone of scattered trees, shrubs and moorland. The shrubs, which include tree heaths, proteas and scattered giant groundsels, grow from a moorland soil covered in grasses, sedges, and herbs. Somewhat about 11,000 feet the taller growth thins out and disappears, leaving an open heath moorland of tussocks, sedges, and low bushes. From some 14,000 feet to the snow line at about 18,000 feet colder and drier conditions give rise to a thinner vegetation of mosses, lichens, and a few semiprostrate herbs and grasses, growing mainly in damp, sheltered cracks between rocks.

Despite the harsh conditions above the forest zone, many animal species survive on the heights. Hundreds of kinds of insect have been recorded on each major peak, and there is a smaller range of reptiles, birds, and mammals. But field observers have to look closely to find insects and other small organisms. Few winged insects risk flying in the light but persistent winds, though many appear when the bushes are shaken and the flowers examined minutely. Similarly the centipedes, scorpions, and other ground-living invertebrates seldom appear in the open; instead they seek out cavities among the rocks and roots to escape the surface extremes of temperature and humidity. The reptiles that feed on these small creatures hunt mainly during the morning and evening when air temperatures are relatively stable. Certain chameleons occur at the upper edge of the forest and occasionally in moorland above 10,000 feet; skinks and other lizards range up to 13,000 feet, living close to lake edges in other situations where insects are plentiful on the ground. Birds of the moorland and alpine zones include plovers, snipes, hill-chats, grass-warblers, and canaries. Among highly specialized feeders are the malachite sunbird, which seeks insect larvae in the rosettes of giant lobelias, and the purple-breasted sunbird, which takes nectar from flowers of the same plants. Alpine swifts are sometimes seen whirling and swooping high over the mountains, probably catching flying insects swept up with warm air rising from the plains and forests below. White-necked ravens scavenge on the ground, and buzzards, augur buzzards, and lammergeiers patrol constantly during the day, buoyed on the updrafts.

Small rodents thriving above timberline level include the tiny harsh-furred mouse, zebra mouse, and four-striped grass mouse. Little is known of how they live, but they lack neither potential food in the form of insects and seeds, nor cover to protect them against the harsh daily range of temperature and the attentions of predators. Groove-toothed rats live high on the upper slopes of Mount Kilimanjaro, nesting among the rocks in cavities that they furnish with sticks and twigs. Their food is mainly soft vegetation, including mosses and new grass, which they collect from the damper upland areas. Dry ground is colonized by mountain hyraxes. These are rabbit-sized mammals with hoofed toes, and they live in large family groups among the high crags and scree slopes of the alpine zone. There are also rock hyraxes, which inhabit Mount Kilimanjaro and many other East African peaks. Mole rats, the size of small guinea pigs, live in good grassland where the soil is deep and rich. They dig deep burrows, throwing up "molehills" with their strong limbs and using their huge orange incisors to cut the grass stems and roots that are their main source of food.

Many kinds of large browsing mammals find their way up to the high moorland pastures from the plains and forest, especially in dry weather when food is scarce below. Mountain reedbuck, bushbuck, waterbuck, eland, bongo, steinbok, and common duiker have all been reported, and elephant, buffalo, and giant hog are frequent visitors from the forests. Klipspringers have been seen up to 13,000 feet on the high crags of Kilimanjaro, playing the role occupied by mountain sheep or goats north of the Sahara Desert. Not surprisingly, these populations of small and large herbivores attract a range of predators. Leopards are universal hunters, taking all sizes of prey from buffalo to hyraxes and small antelope; serval cats and civets have been recorded above 11,500 feet on Mount Kilimanjaro and other mountain moorlands, and lions occasionally follow game mammals up to the high ground when drought strikes the plains. Buzzards, lammergeiers, and Verraux's and tawny eagles hunt by day, Mackinder's and Cape eagle owls by night, taking mole rats and other small mammals on open ground.

The lesser volcanic cones of eastern Africa show similar patterns of zonation, though few reach higher than the zones of moorland or alpine vegetation. Mount Elgon, once perhaps the highest of all the volcanoes but now reduced by weathering and slumping, has a well-developed zone of forest with tall cedars, podocarps, and bamboos. Among its forest animals are flying squirrels, large fruit-eating bats, porcupines,

Dappled light and body spots combine to camouflage a leopard waiting in a tree for prey to pass below. Leopards seek food up through East Africa's mountain forests and on moorland above.

and pangolins—yard-long mammals with scale-like armor that feed almost entirely on ants and termites. Elgon opens on top in a vast caldera or crater with a floor at about 13,000 feet containing a lake and an expanse of thin, alpine vegetation. Cones with summits below 11,500 feet often contain a crater lake and rich grassland, both of which attract herbivores when the plains below are parched and exhausted. The Ngorongoro Crater of northern Tanzania is such an oasis, with a partially forest-carpeted caldera 9 to 12 miles across. Over a dozen species of grazing and browsing mammals live in or visit the caldera, with lions, cheetahs, hyenas, jackals, and hunting dogs as attendant predators.

Leaving the highlands associated with Africa's Eastern Rift Valley and crossing the high plateau of Lake Victoria, we come to the Western Rift Valley. This is the longer arm of the rift system, cutting westward into northwest Uganda then south to form the eastern border of Zaire. Like the Eastern Rift, the valley is marked by a series of mountain ranges, lakes, and the long narrow Lake Malawi. The rift finally disappears in the bed of the Indian Ocean south of the Zambezi River.

Standing between lakes Albert (or Mobuto), Edward (or Idi Amin), and Kivu in the northern part of the Western Rift Valley are Mount Ruwenzori and the Virunga Mountains—two blocks of high mountains of different origin and structure but with similar plants. Ruwenzori is a block of ancient volcanic and crystalline rocks that is about 80 miles long with an average altitude of 10,000 feet. The highest peak is Mount Stanley (also called Mount Ngaliema), over 16,700 feet above sea level. All the high peaks have permanent fields of snow down to about 13,000 feet.

Ruwenzori stands on the border between Zaire and Uganda. The southern tip cuts the equator close to Lake Edward and its northern half rises south of the shore of Lake Albert. This is a region of sweltering heat and persistent rainfall; some 80 inches of rain fall annually on the middle and lower slopes of the mountains, which are thickly forested up to 8500 feet in the east and 10,000 feet in the wetter west. The dense forest of the western slopes arises directly from the lowland rain forest that fills the Congo Basin, and animals pass freely from one kind of forest to the other. In the east, mountain forest rises from drier savanna and parkland. Below about 7500 feet podocarps and other tall hardwoods are the dominant trees, with a cross-lacing of lianas and trailing epiphytes and a gloomy understory of tree ferns and shrubs. Above this level bamboo grows strongly, forming dense, dripping thickets among the taller trees of the forest. From 9000 feet the forest gives way to heathland, with high tree heathers overshadowing a ground cover of thick mosses and shrubs, and open ground

Left: spurwing plover photographed at Lake Nakuru, Kenya. Some kinds of wading bird found by such lakes of the Eastern Rift Valley also occur on the high mountain moorlands of East Africa.

Bamboo forest forms a dense belt above rain forest and below heath on the tall mountains of East Africa. Under ideal conditions this giant grass may reach a height of 100 feet in only 2 months.

Below: a placid crater lake in Uganda's Ruwenzori National Park, located in the Western Rift Valley area. This old volcano reminds us that here, as in Africa's Eastern Rift Valley, many mountains had their birth in molten materials that flowed up through the earth's crust, then cooled to form solid rock.

dominated by large tussocks and rushes. The highest zone of alpine vegetation extends from about 12,500 feet to the snow line. This is a cold, cloudy region of wet moss, small, tightly packed tussock grass and low scrub, with a scattering of tree-groundsels in sheltered corners.

Between Lake Edward and Lake Kivu stand the Virunga Mountains, a group of volcanoes of recent origin that fill a stretch of the Western Rift Valley, and continue southward into the highlands of Rwanda on the eastern shore of Lake Kivu. The western Virungas include the volcanoes Nyamlagira (10,026 feet) and Nyiragongo (11,400 feet). Both have erupted from time to time. The eastern mountains, higher but no longer active, are snow-capped for part of each year; Karisimbi, highest of all, reaches 14,787 feet. Like Ruwenzori, the Virungas arise from dense forest in the west, and they support mountain forest up to 10,000 feet or more.

Both Ruwenzori and the Virunga Mountains lie partly at least within national parks or conservation areas, so their wildlife gets some protection. In the mountain forests of the Ruwenzori can be seen the small forest elephant, which stands only two thirds as tall as the plains elephant, and the dwarf buffalo, a smaller, paler version of the massive lowland forest buffalo. Chimpanzees are plentiful in parts of the forest, up to about 9000 feet; small parties of a dozen or more wander through the trees in search of fruit and fresh leaves, sleeping on platform nests at night. Noisy parties of colobus monkeys are active in the treetops throughout the day, adding their chatter and screams to the cackle of touracos, parrots, and hornbills that generally live at the same level. Looking like agile guinea pigs, tree hyraxes, smaller kin of the rock hyraxes that live in the alpine zone of the eastern mountains, scamper in the trees at night. On Ruwenzori, where there are no rock hyraxes, tree hyraxes extend upward into the alpine zone and live among the crags.

Mountain gorillas range widely in the mountain forests from the equator south to the northern end of Lake Tanganyika. Although absent from Ruwenzori, they are found nearby and in the Virunga Mountains. But their groups are scattered and at most there may be no more than 15,000 individuals in the whole of the forest. These are splendid animals, predominantly black, though large males acquire a silver-gray back on maturity and can weigh up to 400 pounds. Usually led by a mature male, mountain gorillas move in small family bands of 12 to 20, taking most of their food at ground level or in low branches. Adults consume about 50 pounds of food each day, mostly made up of fruit, leaves, and the soft inner tissues of bamboo, wild celery, tree ferns, and other large plants. Many feed almost exclusively in the lowland rain forest of the foothills and rift valley sides, but some range up through the montane forest as high as the bamboo zone. Protected areas have been provided for mountain gorillas, notably around Kikeno and Muhavura in the Virungas, but continuing hunting and pressures to clear the forest place the remaining populations at risk.

The high tablelands of Africa extend westward into Cameroon and the hill regions beyond the Niger, and occupy Africa south of the Congo Basin. In the west the tallest mountains are the long range of volcanic peaks that make up Cameroon's western highlands where Mount Cameroon itself, at over 13,000 feet, is West Africa's highest summit. Rising from tropical rain forest and mangrove swamps, the volcanoes are ringed at successive levels with broad zones of mountain forest, bamboo thickets and wet cloud forest, and open grassland with shrubs and stunted trees. Plants and animals closely resemble those of the East African highlands, despite the distance between them.

Southern Africa's highest mountains are the Great Escarpment and its outliers. This raised edge of the African plateau consists mainly of sandstones capped by volcanic basalts, and it rises dramatically above the coastal plain in a

series of steps. In the southeast the escarpment forms the Transvaal and Natal Drakensberg ranges, which attain over 11,000 feet along the Natal-Lesotho border. The Drakensberg is dry, so that, though soils are fertile, the deeply-scored valleys carry only a thin forest of olive trees, proteas, and other semiarid-land shrubs, while open ground bears grassland and thorn scrub. Hunting has savagely reduced once-plentiful game, and mammal herds are only gradually recovering in protected parks and other sanctuaries. Eland, mountain reedbuck, oribi, gray rhebok, bushbuck, duiker, and klipspringer are still about, though, and there are small groups of jackals and baboons. Lanner falcons, lammergeiers, and black and martial eagles are prominent raptors.

The southwestern escarpment is drier still; overlooking the coastal Namib Desert, it receives moisture mainly from fog, and its arid, wind-carved slopes support grasses and drought-proof shrubs found nowhere else in the world. Rock hyraxes, klipspringers, rock hares, and the rare Hartmann's zebra all somehow survive the harsh drought of this remote region where we now end our brief look at life in Africa's mountains.

Chimpanzee feeding amid the treetops in Zaire's Kahuzi Biega National Park. These primates range up to 9000 feet on the forested lower slopes of the Virunga and Ruwenzori mountains.

The Southern Continents

A wilderness of steep slopes clad in tropical forest makes up much of New Guinea's central mountain spine. Like the other highlands dealt with in this chapter, New Guinea's mountains have life forms unknown in Africa, Eurasia, or North America.

The great northern landmasses of Eurasia and North America are partly balanced in the south by Africa and three other widely spaced continents—South America, Australia, and Antarctica. In structure and geology these three southern landmasses have more in common than their isolation suggests. Indeed, with Arabia, Madagascar, and the Indian subcontinent, all once formed the southern half of the primeval landmass Pangaea—the half that split off as the southern supercontinent Gondwanaland before itself breaking up. South America began to move westward from Africa some 200 million years ago; Australia and Antarctica drifted east and south together for many million years before splitting and shifting to their present positions.

At the time of their dispersal, South America, Australia, and Antarctica probably carried similar plants and animals, even identical species of trees, insects, reptiles, birds, and marsupial mammals. For this reason some of the plants and animals of Australia and South America still relate to and resemble one another. Some Antarctic organisms, too, may have shared these similarities up to 4 or 5 million years ago. Since then, Antarctica's southward drift into polar latitudes has allowed ice caps to cover almost all the continent, wiping out most land-based life.

South America and Antarctica are well endowed with mountains, and on completely different scales and in different ways have developed mountain plants and animals. Australia, though scenically impressive, has few major mountains and few mountain life forms. But New Zealand and New Guinea, which share Australia's corner of the world, are rich in high and relatively new mountains, and in highland plants and animals.

Our mountain survey of these southern lands begins in South America. Upland blocks occur in eastern South America from Venezuela south to Argentina. But far more impressive are the Andes, which closely parallel the western coast. These mountains result from the westward drift of the crustal plate bearing South America. This plate has long been grinding against the edge of the Nazca plate—a neighboring slab of the earth's crust flooring part of the eastern Pacific Ocean. Now the continental plate overrides this oceanic plate, and a deep submarine trench has formed off the western flank of South America. At the same time the clash of plates has brought buckling, folding, and upwelling of molten rock, and it is these events that thrust up the Andes, the world's longest continuous and second-highest mountain system.

The Andes extend some 5500 miles from Staten Island and Tierra del Fuego in the south to the Caribbean shore in the north. For much of their length they consist of two parallel ranges—the Eastern and Western cordilleras. These merge in the far south but are separated in the center by broad plateaus and valleys. The cordilleras fuse in the north-central Andes, then split into four distinct ranges in the far north. The oldest Andean rocks are sediments of mud and sand laid down off the shore of a relatively flat continent in Paleozoic times, 230–450 million years ago. Subsequently folded, faulted, and injected with molten magma from below, these rocks now form the Eastern Cordillera. Volcanic activity between mid-Mesozoic and early Tertiary times raised the eastern mountains by further intrusions of molten rock. Meanwhile, in offshore waters, the same activity raised a double series of mountainous islands that further uplift joined to the mainland where they became the Western Cordillera. Finally, in a bout of volcanic activity beginning 15 million years ago, ashes and lava poured from fissures in the ground, blanketing many of the older ranges, filling the valleys with debris, and building many of the highest peaks of the modern Andes. Many peaks are still volcanically active. Moreover the whole chain of the Andes is subject to earthquakes caused by the continuing friction of plate against plate in the depths of the earth.

Many Andean peaks are impressively high. Numerous northern summits rise above 16,000 feet, while between Ecuador and central Chile a number exceed 20,000 feet. These include Chimborazo in Ecuador, Huascarán in Peru, Sajama in Bolivia, and (in central Chile) Aconcagua—at 22,834 feet the highest peak in the Western Hemisphere. South of central Chile the mountains are lower and few attain 11,000 feet. The snow line rises gradually from about 2000 feet in

the far south to 19,000 feet above the dry Atacama Desert of northern Chile, falling to 16,000 feet in the damper equatorial region.

Since its break with Gondwanaland, South America has been only intermittently connected with other landmasses—possibly with Antarctica through the island chain of the Scotia Arc, certainly with North America through a chain of islands in early- and mid-Tertiary times, and by the Panama land bridge during the past 5 to 10 million years. These geologically brief contacts deeply affected life on the continent, for they let in "new" kinds of creatures and plants, some of which multiplied at the expense of native species. This is largely why only a few kinds of South America's original marsupial mammals remain. Most marsupials live in lowland forest, though some thrive in and well above the mountain forests. The many kinds of placental or "higher" mammals that have entered South America from the north include monkeys, shrews, rodents, carnivores, armadillos, and several kinds of browsing and grazing hoofed animals. Since their arrival, placental mammals have become diversified and now inhabit forests, plains, swamps, and other types of terrain; but while some placentals are basically mountain creatures, few restrict themselves to a narrow altitudinal zone. Birds have traveled more freely than mammals between North and South America, diversifying particularly in the lowland forests of the Amazon Basin but spreading also over the mountains and becoming adapted to local conditions.

Like the Rocky Mountains of North America, the Andes from north to south span a wide range of climatic zones, and are high enough to generate widely differing climates on their eastern and western faces. Southward from about 35°s they include temperate and subpolar mountains, constantly open to heavy rain and snow from the west and chilled by the Antarctic pack ice only a few hundred miles from their southern extremity. In southern Chile, Andean slopes support a dense, somber rain forest, dominated by southern beeches (both deciduous and evergreen) and podocarps, with araucaria pines, tall timber trees, and a dense understory of dripping tree ferns, laurels, and bamboo thickets. Higher

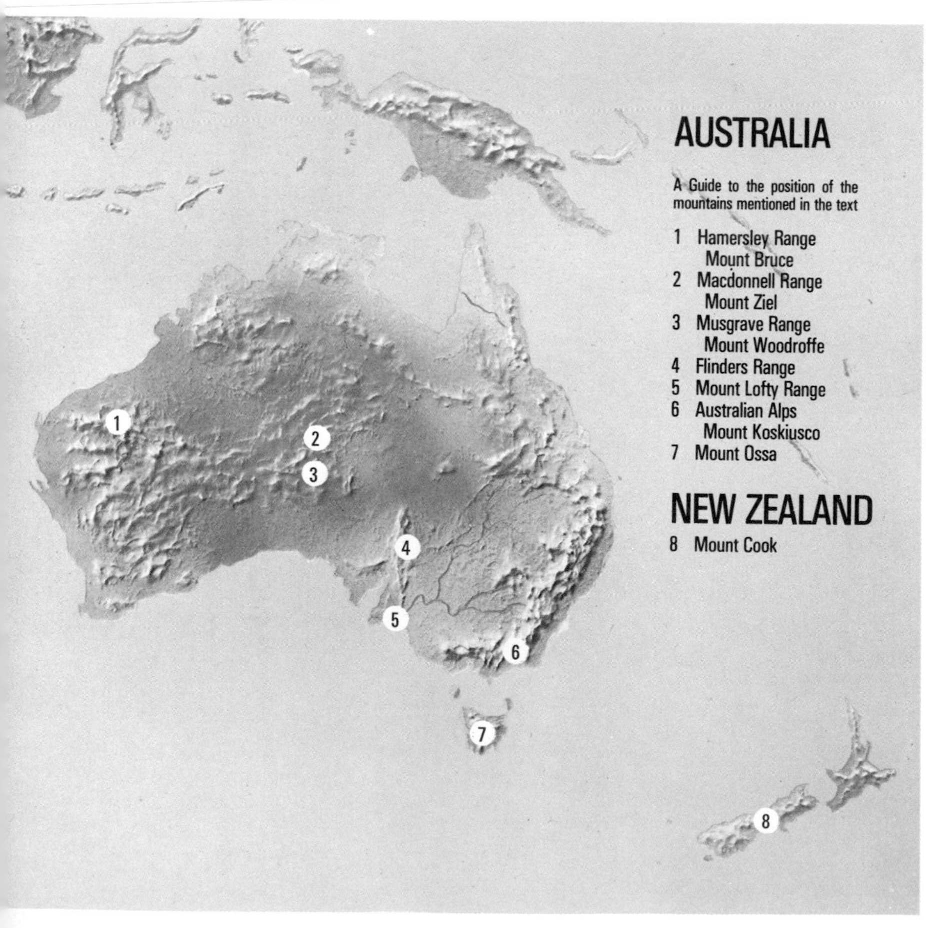

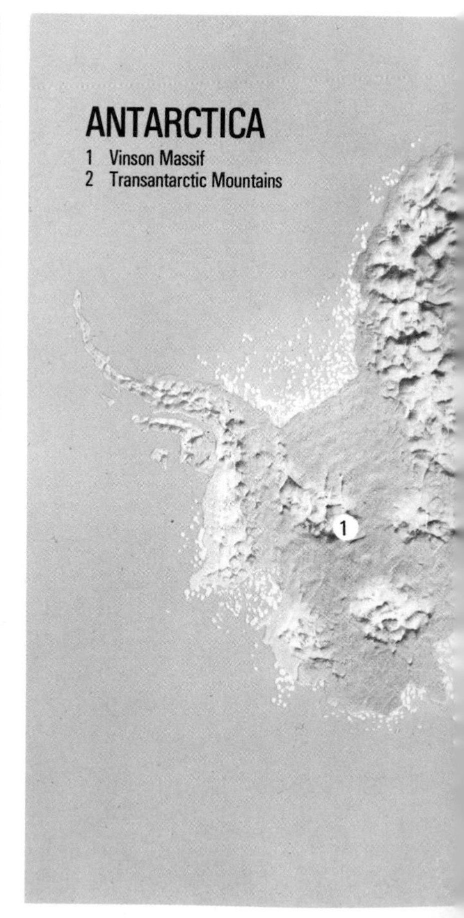

ground and eastern ranges hold drier forests of araucaria, oak, larch, and deciduous southern beech, giving way in places to grassland, scrub, alpine moorland, and even desert. In the far south the mountain forest is gradually replaced at all levels by cold, windswept moorland, dotted with bogs of sphagnum moss and scattered stands of gale-torn southern beech. The highest peaks of this southern section of the Andes carry permanent snow, and many valleys brim with glaciers that sweep down to sea level.

The mountain forests of the southern Andes are somewhat isolated from the rest of the continent by high mountain passes, and support only a few kinds of mammals. Among these are scattered populations of Andean deer that possess simple forked antlers and small tusks; they roam the high forests and grasslands up to 15,000 feet in summer, taking to the warmer valleys in winter. Southern pudu, among the smallest of all deer and weighing only 20 pounds, are more purely forest animals, feeding on mosses and lichens and browsing among shrubs and the lower branches of trees. Small ratlike creatures of the genera *Octodon, Spalacopus,* and *Ctenomys* (known locally as *cercas, ratones topos,* and *tuco-tucos*), burrow extensively in the well-drained soils of grassy slopes and forest ravines, scooping out tunnels between underground chambers where these rodents store roots and tubers for their winter food supply. The dank upland forests shelter several species of water birds, including colorful torrent ducks, and predatory great horned and ferruginous pygmy owls.

North of the 35th parallel, the Andes divides into western and eastern ranges. The Western Cordillera of the central Andes is drier and (except for the highest parts) warmer than most of the southern Andes. In the temperate regions of central Chile the Western Cordillera supports

Mountains of the southern continents: Australia's are the lowest; Antarctica's rise through ice; South America's Andes comprise a long, lofty, unbroken chain.

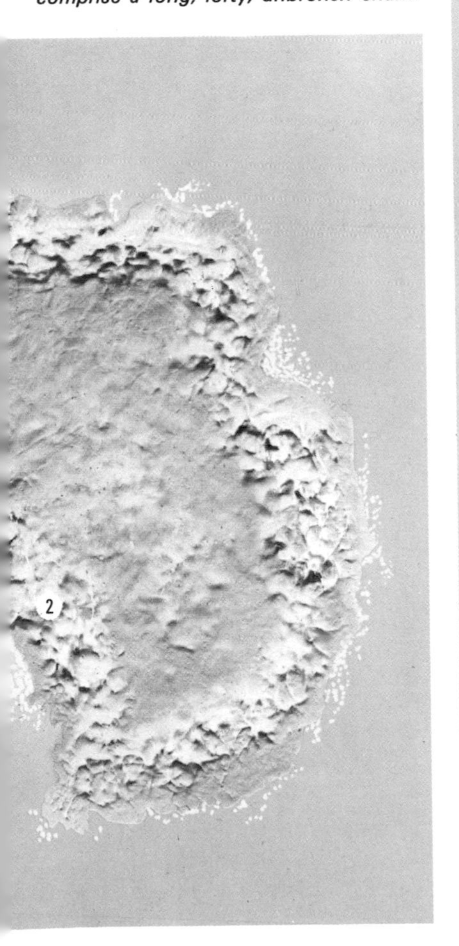

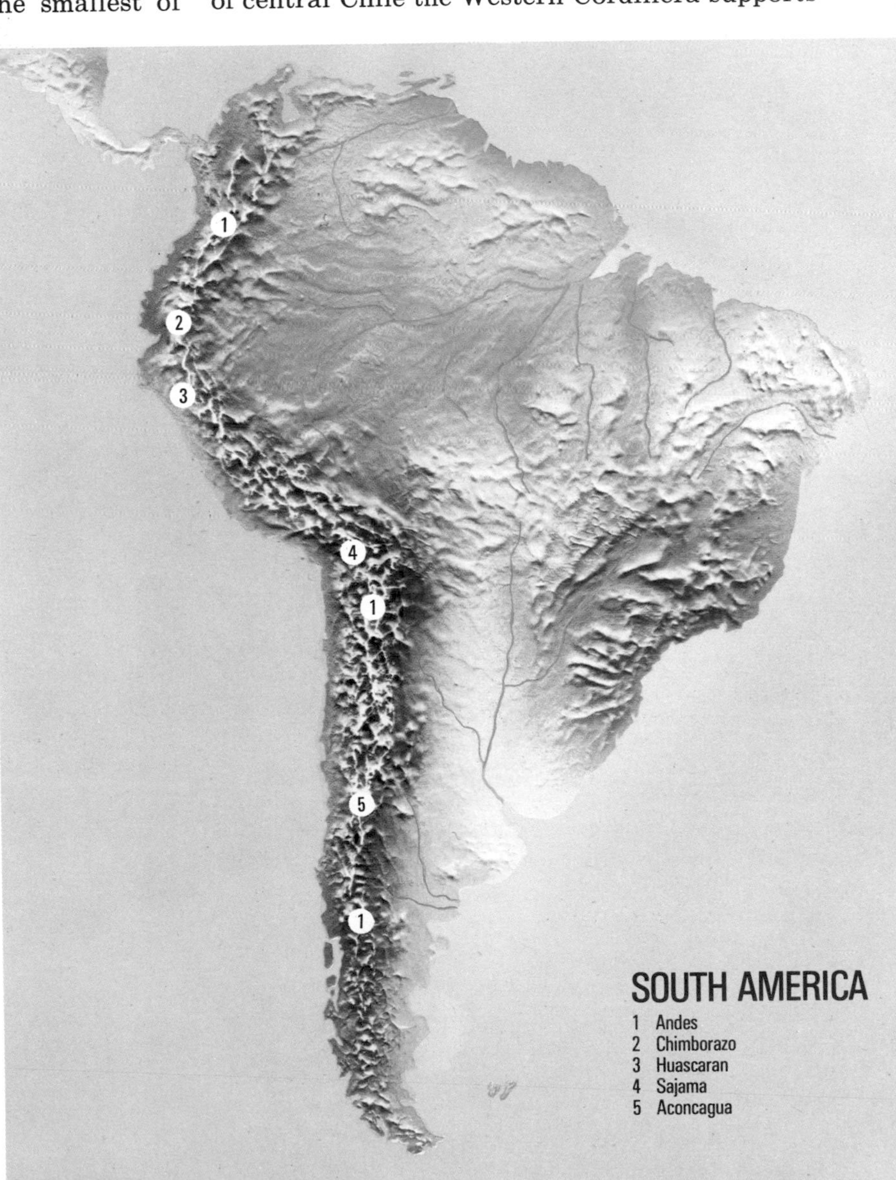

Chilean pines or "monkey puzzles" on a mountain in southern Chile. Their genus, Araucaria, *occurs in South America and Australasia—one proof that these lands were once connected.*

grassland and scrub, but the tropical deserts to the north grow little more than drought-resistant cactus and thornbush. In the driest regions of Chile and Peru the western foothills of the Andes rise from an almost perpetually rainless desert. At heights between about 2000 feet and 3000 feet, cool oceanic air brings seasonal moisture to the mountain slopes, supporting a narrow zone of *loma* or fog vegetation. Above this lies a broad zone of barren semidesert, hot by day and cold by night, with a scattering of eucalypts and other hardy trees on the open slopes, and forests clothing valleys gouged by the torrents that intermittently rush down from the heights above. From about 13,000 feet upward the Western Cordillera carries a so-called *puna* zone of rough grassland and scrub, merging into a narrow band of alpine desert below the line of permanent snow. Between the Eastern and Western cordilleras are valleys and high, rolling hills, with broad plateaus up to about 15,000 feet above sea level. The lower slopes of this area were once forested, but people have largely cleared the trees for fuel and farmland. The puna of higher levels supports grazing mammals, notably the domesticated llama and alpaca (to be described in detail later). Puna covers most of the Altiplano, a rolling plateau that occupies part of eastern Bolivia, separates the outer Andean ranges and includes the large lakes Poopó and Titicaca. Despite its position in the tropics, high altitude makes the Altiplano a cold, windswept plain, with frequent night frosts. Even so, the plateau is densely peopled, containing many large settlements, such as Oruru and Puno, and one city—La Paz—of over 800,000 inhabitants.

The central and northern Andes have attracted a wide range of mammals from the lowlands. Several species of tree-living marsupial dwell in the mountain forests, where they feed mainly on fruit and insects. Mouse or murine opossums—small, shrewlike animals with long, prehensile tails—occur up through the cloud-forest zone. These pouchless marsupials carry their young on their back and flanks. *Colocolos*, closely related but pouched, live in the mountain forests of Chile. Rat opossums, larger, pouchless, and with long, non-grasping tails, are common both in high forests and above the timberline; in the Peruvian Andes they flourish well above 14,000 feet.

Among the placental mammals, few species of shrew managed to invade South America, and these only reached the northwest where two species of the genus *Cryptotes* live in the highlands of Ecuador and Colombia. But South America is remarkable for the variety of its rodents, the much diversified descendants of those that invaded from the north at different times during the late Tertiary period. Many have taken successfully to mountain living in at least some part of the Andes. Squirrels are curiously rare, though two species of the genus *Leptosciurus* live in the high forests of Colombia. Species of mouselike rodents at home in the highlands

Patagonian armadillo, a placental or "higher" mammal. After land linked the Americas, higher mammals entering South America ousted many of its native marsupials.

The Eastern Cordillera of the central Andes and the four ranges of the northern Andes all rise entirely or partly from tropical forest. In the east this is the rich rain forest of the Amazon Basin, with its astonishing variety of trees, shrubs, lianas, and epiphytes and its wealth of birds, insects, and mammals. This dazzling array of tropical life extends up the foothills to about 5000 feet. Above is a narrow band of cloud forest, permanently soaked by condensed moisture carried west by the trade winds. At 10,000 feet or more the trees are dwarfed and gnarled by the wind, forming a broad zone of "elfin woodland" that opens onto puna grassland and alpine desert in the colder, drier zones above. The higher alpine desert has stands of tussock grasses and herbs alternating with patches of cactus 9–12 feet high and squat rosette plants with tall flowering spikes, resembling the giant plant growths of the East African highlands.

Indian reed boats on Titicaca, the largest lake in South America. Coots, cormorants, ducks, geese, grebes, gulls, plovers, and snipe live here, more than 12,000 feet up in the Andes.

above 10,000 feet include large-eared *pericots*, silky *chinchillulas*, burrowing mole-mice *(Notiomys)*, and volelike mice of the genera *Akodon* and *Punomys*; these are mostly creatures of the drier grasslands and forests, and they feed on insects, seeds, seedlings, shoots, and carrion. The Altiplano and other big grassland areas support populations of Andean rats or *chazcharitos*, and several kinds of larger rodents related to the guinea pigs and porcupines that figure strongly among South American mammals. Guinea pigs themselves are widespread in the Andes, where peasants raise them for food. Tree porcupines or *coendus* of several kinds are found in the mountain forests, one species reaching 8000 feet in the highlands of Ecuador and Colombia. These carry a formidable armory of quills, and are almost entirely arboreal and vegetarian. *Maras*—long-legged guinea pigs weighing up to 30 pounds—live like very large hares in the uplands of southern and central Argentina; *pacaranas*, of similar size, are short-legged and heavily built rodents of the northeastern Andes where they browse peaceably in the high forests but are hunted ruthlessly by man for their tender meat.

Chinchillas once lived in large colonies at all levels of the Andes up to 20,000 feet above sea level. About the size of a small rabbit, but with long, silky gray fur, they have been hunted almost to extinction by generations of trappers, and are now very rare in areas accessible to man. Chinchillas are usually at home among rocks and rough ground, burrowing for warmth and protection; those that live highest in the mountains tend to have the longest fur and the shortest ears and tail, safeguards against chilling and frostbite. Mountain viscachas, closely related to chinchillas, are larger animals with dense gray fur that—fortunately for them—has little commercial value. The three Andean species live high in the grasslands and desert regions from Bolivia south to Patagonia; from afar they resemble large, portly rabbits with long tails carried high and curling upward at the tip. Viscachas live in rather dry, desolate regions,

Andean Indian women (one spinning wool). Andean Indians can work hard in rarified, high-altitude air. Surviving severe oxygen lack, some live semipermanently above 20,000 feet.

usually in small colonies centered on dens—shallow burrows or cavities among rocks. Intensely social, they sunbathe in groups early in the mornings and huddle together for warmth; their food is mosses, lichens, and tough mountain grasses and shrubs, and they communicate by high-pitched whistles that carry well on the mountain air. *Cercas, ratones topos*, and *tuco-tucos*—widespread in the southern Andes—also thrive on high ground in the north where they burrow communally like gophers in soft soil. Chinchilla rats—again more closely allied to guinea pigs than to rats—live in the central Andes of Peru, Argentina, and Chile, burrowing among rocks at elevations of up to 8000 feet. Mountain *pacas* are larger creatures of burrowing habit, colonizing forest and open ground 6500–13,000 feet above sea level. Their powerful hind limbs, armed with sharp claws, dig the holes that protect them from weather and predators. Huge orange incisors serve for gnawing stems and roots, and unusually broad cheek bones accommodate pouches that resonate to provide the pacas with a repertoire of sounds—whistles, grunts and hollow groans.

Neither goats nor mountain sheep have found their way south down the Panama isthmus. Their niche in the South American highlands is taken instead by four humpless species of the camel family that browse and graze along the length of the Andes wherever their meager food needs are met. The largest wild species is the guanaco, weighing up to 200 pounds and standing as tall as a man. Guanacos feed mainly on grasses and shrubs and wander in small, compact groups, usually on open plains where their would-be predators cannot lurk unseen. Widespread at various levels, they even occur in semidesert 13,000 feet above sea level. Vicunas look like smaller versions of guanacos but weigh only half as much, and seldom live below 11,500 feet; their haunts are the steeper slopes of the mountains between timberline and permanent snow. Often they feed in territorial groups, each comprising a leading male and several females with their young. Larger bands of unattached

males also wander the heights, grazing throughout the day. Men hunt guanacos and vicunas for their meat and wool and in parts of the Andes periodically round them up for shearing. The third and fourth species of camelid are the domesticated llama and alpaca, derived from guanaco and vicuna stocks by selective breeding or (as seems more likely) from species now extinct in the wild. Llamas are large animals, weighing up to 300 pounds. They serve for carrying packs and as a source of meat, hides, and warm but coarse wool. Alpacas are smaller and kept chiefly for their clip of fine silky wool, which both alpacas and men value for its lightness and warmth.

The high forest has many species of mammals peculiar to South America. Tapirs, related both to rhinoceroses and to horses, mainly inhabit the northern and eastern tropical regions. The Brazilian species, restricted almost entirely to wet forest, roams the foothills. The mountain tapir, equipped with a woolly coat, lives at 10,000–13,000 feet in the Andes of Ecuador, Colombia, and Peru. As tall as a small pony, but much heavier in build, tapirs browse and graze in dense thickets of the forest, using their short, mobile trunks as sensors to find tasty, mainly vegetable, food among the forest floor debris. Andean deer browse up to 14,500 feet on the central Andes as far north as Ecuador; through northern Ecuador and Colombia a northern species of pudu ranges the highest forests, while brockets—another group of small red deer with single-spike antlers—are restricted mainly to the lower slopes of the mountains.

Coatis and kinkajous—arboreal mammals related to the raccoons of North America—feed by day and night in the canopy of the northern forests, ranging high up into the mountains. Coatis are black, red, or dark tan beasts the size of small terriers, with ringed tails longer than their bodies and usually held high like a banner. They move through the forest in small bands, snuffling with their long noses in crevices and leaf litter as they seek insects, millipedes, spiders, small mammals, reptiles, and fruit, both in the trees and on the forest floor. Females and their young live together throughout the year; males join the bands only for breeding. Kinkajous are smaller animals, almost entirely arboreal and mainly vegetarian, with yellow-buff fur and long prehensile tails; usually nocturnal, they wander the forest in pairs or small family groups, gathering their year-round harvest of fruit from the trees and bushes. Mantled and red howler monkeys, spider monkeys, and other primates feeding mainly on fruit, leaves, and shoots, are noisy competitors in the lower mountain forests of Ecuador and Colombia. South America's only species of bear—the spectacled or Andean bear—also occurs in the northwestern mountains. Black or dark brown, with distinctive white markings on the face, throat, and chest, it feeds mainly on leaves and shoots, sometimes climbing tall trees or breaking saplings to reach succulent food. Spectacled bears spend much of their lives on the high grasslands, but descend to the valley forests and plantations as well.

South America has its own predatory mammals, some of which follow their prey onto the higher

Left: tree porcupine, member of a prehensile-tailed genus of porcupines that extends from south Mexico through tropical South America. Several species live in the Andean forests, where one occurs 8000 feet up near the equator.

Right: Ezseletia moritziana *photographed in Venezuela. Such plants thrive on Andean alpine deserts, their rosette shapes, adapting them to high mountain life near the equator, are similar to those of their counterparts in East Africa.*

slopes of the mountains. Most versatile are the pumas or mountain lions—large, unspotted cats with yellow, tan, or black fur, belonging to the same species as the pumas of the Rocky Mountains. Large males weigh 220 pounds or more; females are smaller, but still able to stalk, kill, and dismember the larger herbivores that are their main prey. Pumas range widely through forest, grassland, and desert, stalking from cover and knocking over their victims with a final rush. Prey species include deer, young tapirs, guanacos, vicunas, and the llamas, alpacas, and other domestic stock of mountain farms. Pumas are accordingly unpopular in the densely inhabited grasslands of the Andes, where men try to hunt them out of existence. Jaguars are slightly larger, much heavier cats, camouflaged in brownish yellow with black spots arranged in rosettes. Though mostly animals of the forest, they are open-minded in their approach to hunting, emerging readily into open country and seeking large and medium-sized herbivores over desert, scrub, and grassland. They swim well and catch fish with their paws (reputedly using their tails as lures), disembowel alligators, and dig reptile eggs from nests on the sandbanks. Smaller predatory cats include the ocelot, the delicate margay, Andean cat, and lithe, otterlike jaguarundi, which feed mainly on birds, reptiles, monkeys, rodents, and other small animals both

Alpacas pose docilely near the remains of a fortress built by Incas, an Andean Indian people. Andean Indians probably derived alpacas from wild cameloid ancestors now extinct.

on open ground and in the forests. Neither large nor small cats are numerous, but both are found up to about 10,000 feet through the central and northern Andes. Maned wolves—long-shanked foxy creatures with a shaggy red-tan coat and pointed ears—are more plentiful on the high plains, though solitary and wide-ranging. They dig reptiles and small rodents from their burrows, and run down larger mammals. Crab-eating foxes, smaller and stockier, hunt burrowing mammals and ground-nesting birds, and both species reportedly vary their diet with fruit and berries.

South America has an extraordinary wealth of bird life. Colombia and Ecuador alone outrank all North America in number of species, most of them found in no other continent. Many birds are adapted for life in the lowland forest, but the mountains have their share of this rich avifauna; their forested zones are full of lively, colorful birds, and the high grasslands of the central and northern Andes have attracted many species from lowland plains of the south, where living conditions are similar. Cloud forest perhaps has the richest variety, with vivid hummingbirds, trogons, cotingas (including the brilliant red-bronze cock-of-the-rock, and curiously wattled bell-birds with penetrating, monotonous calls), and brilliant motmots, tanagers, and jacamars. The high, damp grasslands of Colombia and the drier puna of Peru and Bolivia have many ground-nesting birds of more sober plumage—ovenbirds with mosquelike domed nests, earth-creepers, ground tyrants, Andean flickers, tinamous, and a host of thrushes, wrens, and finches that feed on the abundance of insects, seeds, and young shoots. Several species of hummingbirds, including insectivorous hillstars, live in the high grasslands, building their tiny hanging nests on cliff faces that catch the morning sun. The high-country lakes attract many species of water birds; there are several native ducks, coots (including a giant species as big as geese), flightless grebes, and cormorants, and high saline lakes support populations of flamingos. There is even an Andean species of gull, living close to fresh water 13,000 feet up. Most spectacular of Andean birds are the predatory *urubus* (vultures) and condors, which circle in the sky at heights well above 16,000 feet, swooping to feed on lesser creatures and carrion on the mountain slopes below.

Far below the Andean condors' Pacific Ocean horizon stands the island continent of Australia. Continental Australia rests centrally on a massive crustal plate that also bears India in the remote northwest, New Guinea in the North, and New Zealand in the southeast. Shifting slowly northward, the Indo-Australian plate presses hard against the southeastern corner of the vast Eurasian plate, and in turn suffers pressure from the Pacific plate on its eastern and northern borders. The complex mountain-building activities at the boundaries of these plates leave

Blue-crowned motmot, a bird of the tropical forests on eastern slopes of the Andes. Motmots eat snails, lizards, larger insects, and fruit. Most nest in banks in tunnels up to six feet long.

Australia itself relatively unmoved. But besides throwing up the Himalaya and mountains of Southeast Asia, these disturbances created the high folded, faulted, and volcanic mountains of New Guinea and New Zealand, and the many island arcs of the southwestern Pacific and northeastern Indian oceans.

Australia is made up of a high western platform of ancient rocks topped by limestones and sandstones, a broad central basin seamed with usually dry river and lake beds, and an eastern range of modest, mature mountains, raised long ago in the Paleozoic era, worn down then re-uplifted by earth movements and volcanic activity in mid-Tertiary times. Eons of weathering and erosion—latterly in a desert climate—have leveled the faulted and broken surface of the western plateau, making central and western Australia a land of arid sandy plains scattered starkly with rolling hills and bluffs of granite, red sandstone, limestone, laterite, and ancient lavas. The highest peaks of these desert ranges are Mount Bruce (4024 feet) in the Hamersley Range of the far west, and the slightly higher mounts Zeil and Woodroffe of the Macdonnell and Musgrave ranges in central Australia. Though often strikingly colorful and the home of interesting plants and animals, the arid hills of the center bear vegetation similar to that of the plains. Mulga (acacia) scrub grows on dry land (though not on the driest) irrespective of altitude. The deep gorges and valleys of the sandstone hills sometimes accumulate enough water to support a richer growth of mulga, casuarines or desert oaks, tea trees, cane grasses, spinifex, and white-barked ghost gums that hold resident populations of geckos, skinks, snakes, and small marsupials. Most feed on insects or on each other, hunting during the coolness of night. Occasional rain showers clothe the mountains briefly in quick-growing plants that mature, seed, and die within a few weeks. Their organic remains, dry but edible, form the basis of food chains for many months. In southern Australia mallee (eucalyptus) scrub and thin, open forest cover the dry sandstone ridges of the Flinders and Mount Lofty ranges, supporting meager populations of gray and hill kangaroos, koalas, and other browsing and grazing mammals.

The Great Dividing Range, which includes all of Australia's tallest and most spectacular mountains, extends down eastern Australia from Cape York in the extreme north to Tasmania in the far south. Ancient folding, followed by erosion and more recent block-faulting, uplift, and volcanic activity make this a complex range, with high sloping tablelands, steep escarpments and bluffs, and deeply incised valleys with torrents and waterfalls. The eastern faces of the range are in many places precipitous, and well watered with rain and condensation from Pacific Ocean air masses; western slopes tend to be gentler and drier. Many peaks of the range reach 5000 feet or more, but only the highest massifs of the south—those building up to Mount Kosciusko in the Australian Alps and Mount Ossa in Tasmania—are snowcapped for more than a few weeks in winter. Forest covers the mountains from sea level to about 6000 feet. In dry areas tall eucalyptus trees and cypress pines predominate, with a dense understory of shrubs and grasses; where rainfall is heavy and the atmosphere damp, the forest is thicker and more luxuriant. In the south several species of southern beech are dominant, and there are "mountain ash" and other eucalypts. The north has a richer and more tropical rain forest, with huge buttressed trees and shrubs of many species—buoyong, red cedar, flame-trees, palms, eucalypts and wattles, grevilleas, waratah—and a tangled understory of lianas, climbing vines, epiphytic ferns, and dazzling orchids. Above stands a windswept parkland of snow gums and other hardy species; beyond the timberline the high plateaus and peaks carry alpine moorland, with tussock grasses and flowering herbs, and barer, drier rock-strewn heaths. Though there was some glaciation during the Ice Age, these mountains lack the rugged crags of the Alps, Andes, and Rockies, and are almost completely covered with soil and vegetation to their highest peaks.

Within the mountain forest itself zonation effects are slight, and many reptiles, birds, and mammals live with equal success on low ground and high. The egg-laying platypus or duck-bill occurs in mountain streams of the east, where it forms long burrows in muddy banks and swims in search of insect larvae, crustaceans, and mollusks; once hunted for its soft fur, it is now rigorously protected throughout its range. Pouched rats, marsupial mice, and long-nosed and short-nosed bandicoots feed on the worms and insects of the forest floor. Many other small marsupials are arboreal, running like mice or small squirrels along the branches of the trees. Tiny pygmy opossums—among the smallest of all

Tropical forest clings to the sides of this gorge on the hot, humid, eastern flank of the Peruvian Andes. The drier western flank of the Peruvian Andes bears only desert and scrub.

pouched mammals—feed on the nectar and pollen of eucalyptus blossoms. Pygmy gliders, sugar gliders, and other species with patagia or gliding surfaces along their flanks launch themselves from tree to tree, widening their search for insects, flowers, and juicy shoots. Like most other small mammals of the Australian bush they are nocturnal, emerging in the evenings to glide like irresolute bats among the treetops. *Burromys*, another mouselike creature of the high forest, was described originally from fossil remains and thought to be extinct, but has recently been discovered alive in the mountains of the southeast.

Koalas, chubby and bearlike, and brush-tailed opossums are larger arboreal marsupials, which clamber more phlegmatically among the branches munching shoots and leaves. The larger ground-living species include wombats—like stout marmots the size of small pigs—which burrow in the soil and feed on roots and grasses, and gray or forester kangaroos, which browse and graze on the open floor of the forest. Euros or hill kangaroos are often found on the higher, more open parklands of the forest, and rock wallabies have developed a particular facility for feeding, like bouncy mountain sheep, on rough rocky ground. Predatory dogs called dingos (not marsupials but placental mammals introduced by early man) and dasyures or "native cats" with spotted coats are found in the forests of the mainland. Tasmanian uplands harbor two species no longer found in Australia itself—the stocky Tasmanian devil and houndlike thylacine or Tasmanian wolf. Like the dasyure these are true marsupials; both are nocturnal, feeding on smaller mammals, reptiles, and birds. Their fondness for lambs and poultry made them unpopular with early settlers, who hunted the Tasmanian devil off their lands and have probably exterminated the thylacine.

Amphibians and reptiles are common throughout the high forest, and include both tree-living and ground-living frogs that hatch their eggs and complete their life cycles in damp moss. Colorful carpet snakes, some up to 15 feet long, and smaller death adders and other venomous species hunt birds and small mammals on the forest floor. The many species of birds that live in this rich and varied habitat include high-flying parrots and parakeets, graceful lyrebirds, and several kinds of bowerbird that build stages or grottos and decorate them with paint, pebbles, and feathers to attract their mates.

North of Australia stands New Guinea, a tropical island over 1200 miles long with a broad southern plain and a spine of high, cloud-capped mountains along its major axis. Forming part of the edge of a crustal plate, it has long been a region of mountain-building; its steep fold mountains, new and unstable, rise in serrated ridges to 10,000 feet and more in the center and west of the island. The highest stand above 16,000 feet, carrying a year-round mantle of snow and ice from about 13,800 feet upward. Night-frosts above and heavy tropical rains below carve steep gullies in the mountain slopes, which are heavily forested to a timberline at about 11,500 feet. Dense rain forests clothe the lower slopes, to be succeeded at 4000 feet or so by mixed forests of evergreen oak, podocarps, southern beeches, palms, and tree ferns, with an immense variety of other species. At higher levels southern beech predominates, with bamboo, breadfruit, araucaria pines, and palms. Above 9000 feet this mixture develops into cloud forest, mantled with mosses and ferns. Higher still, cloud forest gives way to a narrow zone of dwarf trees and shrubs, dominated by rhododendrons, olearias, coprosmas, and other familiar southern species. The highest levels below the snow line are carpeted with tussock and heathland vegetation. Mountain-forest animals remain largely unstudied, but we know that because New Guinea and Australia were often linked by land in the past, many of their birds and marsupial mammals are closely related.

New Zealand's twin islands lie some 1200 miles from Australia at the southeastern edge of the Indo-Australian crustal plate. Though earlier landmasses stood here, the alpine ranges of the South Island and volcanic massifs of the North Island are relatively new. New Zealand's mountains were heavily ice-capped and glaciated during the Ice Age, and those of the South Island still support permanent snowfields and remnant glaciers. Forests extend to a timberline at about 5000 feet, and tussock grasslands and alpine herbfields continue to 6500 feet or above, grading into a zone of bare rock and snow. Mount Cook, the highest peak, reaches 12,349 feet.

Because of its isolation New Zealand is much poorer in native plants and animals than Australia and New Guinea. There are many kinds of insects, millipedes, spiders, and other invertebrates, but few amphibians or reptiles and no native land mammals. The cool mountain forests, though lush and thickly carpeted with ferns and moss, are dominated by relatively few species of southern beeches, podocarps, tree-ferns, and shrubs; the high grasslands and herbfields lack the variety of flowering species that color European and North American alpine habitats, and even the birds of the high country are monotonously drab.

But man has done much to alter this picture since he first arrived in New Zealand perhaps 1000 years ago. Early settlers, of Polynesian origin, locally burned bush to grow crops, but did little damage to highland flora and fauna. However, European settlers arriving from the mid-19th century onward made profound and lasting changes. By bringing in sheep and cattle, and clearing the forests for pasture and timber, they turned hillside woods into poor tussock meadows. Just as devastating was man's introduction into the high country of game animals from Asia, Europe, America, and Australia. Rabbits, hares, wallabies, opossums, chamois, tahr, and many species of deer, multiplying in the absence of natural predators, have nibbled and browsed their way through the high forests and grasslands, clearing undergrowth, preventing regeneration, opening the canopy, and exposing steep mountain slopes to erosion and slip.

Koalas, like many other Australian mammals, live in both lowland and mountain forest. Most Australian mountains are too low to support distinctive life zones at different levels.

New Zealand's mountain wildlife, then, has become a curious mixture of native and alien species. Native alpine birds include a falcon, now not very common, a rock wren, which hunts busily for insects among the high screes, a pipit of open moorland, and a parrot—the kea—which lives close to the snow line in the high mountains of the South Island. A second mountain parrot—the kakapo was once widespread in high beech forests but now inhabits only one watershed of Fiordland in the Southwest and the wet forested slopes of Stewart Island in the far south. Blue ducks bob about on mountain torrents, and pied stilts, oyster-catchers, wrybills, and other waders breed on high shingle beds and moraines. Black-backed gulls, largest and most enterprising of New Zealand's gulls, nest above 5000 feet near rivers on high pastures, and two other seabird species, Hutton's shearwater and the black petrel, breed on the tops of sea-facing South Island mountains. Many native mountain birds have been displaced by European species shipped in by settlers to relieve nostalgia and to fight insect plagues. Walking in the high country of New Zealand today, you are as likely to hear European blackbirds, skylarks, redpolls, and mallards as the more elusive and quieter native species. Although the introduced mammals are locally plentiful, hunting pressure has made them wary of man. Only sheep are readily seen on a high-country walk, and sheep probably do more damage to New Zealand's mountain slopes and native wildlife than all other introduced species.

Some 1500 miles south of New Zealand is Antarctica. Largest of the southern landmasses,

this continent resembles a vast white comma with tail pointing toward South America. The white surface is in fact a mantle of ice, overlying two land areas. Greater Antarctica, the larger sector surrounding the south geographic pole, has a raised edge forming the Transantarctic Mountains. These rise clear of the ice in Victoria Land, near the coast, but disappear under the ice within a few hundred miles of the pole. Lesser Antarctica, the Pacific side of the continent, is a complex of mountainous islands of mainly volcanic origin, tailing into the long, broken Antarctic Peninsula. Of the southernmost islands only the peaks poke up from the ice, as isolated nunataks and mountain ranges. Antarctica's highest peak is the Vinson Massif, which reaches 16,860 feet in Lesser Antarctica; many neighboring peaks, and others in the faulted blocks of the Transantarctic Mountains, exceed 13,000 feet.

Life on Antarctica's mountains is harshly curbed by constant aridity and cold—summer temperatures seldom reach freezing point and winter temperatures descend well below −60°F. for weeks on end. Yet, astonishingly, the polar mountains support tiny life forms. Naturalists have found mosses, lichens, and algae growing about 6000 feet up near the South Pole itself. Within 400 miles of the pole there are small,

Left: cloud-crowned rain forest covers Mount Réaumur on tiny Long Island, off north-eastern New Guinea near the equator. Below: Jersey cows in fields below Mount Egmont in New Zealand's North Island. Imported animals have helped to transform much of New Zealand's mountain forest into poor pasture.

sheltered patches of vegetation containing insects and mites. These patches thaw and warm in the sun even though nearby air remains well below freezing. The warmth lasts only a few hours at a time, but it is enough to let the plants grow, and the insects and mites feed, mate, and lay eggs. Some mountains close to the sea support a denser growth of mosses and algae, fertilized by droppings from small colonies of Antarctic skuas, terns, and petrels.

Antarctica's insignificant mountain life forms will persist for the foreseeable future. But almost everywhere else in the world, highland plants and animals are threatened by man. To satisfy the needs of his expanding populations, farmers, lumberjacks, and miners plunder mountain soils, forests, and rocks at an increasing rate, sometimes destroying whole wildlife communities in the process. Even man's search for beauty and tranquillity among the peaks carries a threat where thousands of visitors trample and accidentally burn. Man needs ingenuity and sensitivity if he is to protect mountain communities from himself—tough and resilient though these communities have proved in their never-ending struggle with the elements. This is the problem that man must solve before mountain life can be declared safe for future generations to enjoy.

Index

Page numbers in *italics* refer to illustrations or captions to illustrations.

Picture Credits

Key to position of picture on page: (B) bottom, (C) center, (L) left, (R) right, (T) top; hence (BR) bottom right, (CL) center left, etc.

Cover: K. Biedermann/ZEFA
Title page: F. Henrion/Explorer
Contents: John Moss/Colorific
9 R. N. Mariscal/Bruce Coleman Inc.
10 G. F. Allan/Bruce Coleman Inc.
12(B) Heather Angel
13 Tom McHugh/Photo Researchers Inc.
14 Reproduced with the permission of the Royal Geographical Society
15 Urve Kuusik/Colorific
16 C. Weaver/Ardea, London
17 *Daily Telegraph* Colour Library
18 Bill Leimbach
19 Tom McHugh/Photo Researchers Inc.
21(T) Nardin/Jacana
21(B) Heather Angel
22 F. Walther/ZEFA
23 Bryan L. Sage
24(L) Noble Proctor/Photo Researchers Inc.
24(TR) H. Eisenbeiss
24(BR) Weisbecker/Explorer
26–7 J. Thomas/Explorer
29 V. Englebert/Photo Researchers Inc.
30(TL) Lowell J. Georgia/Photo Researchers Inc.
30(B) G. R. Roberts, Nelson, New Zealand
31 Giorgio Gualco/Bruce Coleman Inc.
34(TL) David Moore/Colorific
35(T) Spectrum Colour Library
35(B) B. Thomas/Explorer
36 Nicholas DeVore III/Bruce Coleman Inc.
39 Spectrum Colour Library
40 Lowell J. Georgia/Photo Researchers Inc
41 Chas. Ott/Photo Researchers Inc.
42 C. Mundt/Tom Stack & Associates
43 Jen & Des Bartlett/Bruce Coleman Ltd.
44–5(T) Pissavini/Jacana
44–5(B) Chaumeton/Jacana
46 Peter Steyn/Ardea, London
47 R. Everts/ZEFA
49 Jim Amos/Photo Researchers Inc.
50(R) Bill Leimbach
51 Warren Garst/Tom Stack & Associates
52(L) Peter Miller/Photo Researchers Inc.
53 Dean Brown
54 Myron Wood/Photo Researchers
57(T) John Lewis Stage/Photo Researchers Inc.
57(BL) C. Mundt/Tom Stack & Associates
57(BR) Larry West/Frank Lane
58 Dean Brown
59 Tom McHugh/Photo Researchers Inc.
60 Martin Bruce/Tom Stack & Associates
61 Philip H. Evans
62(T) B. L. Sage/Ardea, London
62(B) Charles R. Belinky/Photo Researchers Inc.
63(L) Charlie Ott/Photo Researchers Inc
63(R) Dupont/Jacana
65 Lowell J. Georgia/Photo Researchers Inc.
68 Nick Holland/Susan Griggs
69 George Hall/Susan Griggs
70 Frances Current/Photo Researchers Inc.
71 Wildlife Unlimited/Tom McHugh/ Photo Researchers Inc.
72(T) Porterfield-Chickering/Photo Researchers Inc.
72(B) C. J. Ott/Bruce Coleman Ltd.
73 Joy Spurr/Bruce Coleman Inc.
74 Lowell J. Georgia/Photo Researchers Inc.
75(TR)(B) Victor Englebert/Photo Researchers Inc.
76 Dean Brown
78(B) C. J. Ott/Bruce Coleman Ltd.
79 Dean Brown
80 G. C. Kelley/Photo Researchers Inc.
82(T) Earl Roberge/Photo Researchers Inc.
82(BL) Dean Brown
83(T) L. Lee Rue/Bruce Coleman Ltd.
83(B), 84 Tom Myers/Tom Stack & Associates
87 Christopher G. Knight/Photo Researchers Inc.
91 Strobino/Jacana
92 Schraml/Jacana
93(T) Heather Angel
93(B) John D. Drysdale/Colorific
94 E. Boubat/Photo Researchers Inc.
96 Thomas D. W. Friedmann/Photo Researchers Inc.
97 Emil Muench/Photo Researchers Inc.
98(T) G. D. Plage/Bruce Coleman Ltd.
99 Chris Bonington/Bruce Coleman Ltd.
100(L) George Holton/Photo Researchers Inc.
101 G. D. Plage/Bruce Coleman Ltd
103 George Holton/Photo Researchers Inc.
105 V. Englebert/Photo Researchers Inc.
108 R. Everts/ZEFA
109 Fritz/Jacana
110 Heather Angel
111 Carl Purcell/Photo Researchers Inc.
112 Dr. L. McCarthy/Tom Stack & Associates
113 Brian Boyd/Colorific
115 C. Weaver/Ardea, London
116 I. R. Beames/Ardea, London
117 J. Van Wormer/Bruce Coleman Ltd.
118(T) A. J. Deane/Bruce Coleman Ltd
118(B) P. Forge/Explorer
119(R) K. Paysan/ZEFA
121 G. D. Plage/Bruce Coleman Lt
123 Jack Fields/Photo Researchers Inc.
126 Photo Researchers Inc.
127(B) Roberto Bunge/Ardea, London
128 Victor Englebert/Susan Grigg
129 Reflejo/Susan Griggs
130 Patricia Caulfield/Photo Researchers Inc.
131 Richard B. Peacock/Photo Researchers Inc.
132 Victor Englebert/Susan Griggs
134 Castel/Jacana
135 Reflejo/Susan Griggs
137 Bill Angove/Colorific
138 Eric Lindgren/Ardea, London
139(R) David Moore/Colorific

Artist Credits

Susan Cook 32(L); David Nockels 33 (Visualized by Michael Turner); Brian Watson 50(L), 66, 89, 106, 124–5